John Call Dalton, Robert James Mann

A Guide to the Knowledge of Life, Vegetable and Animal

John Call Dalton, Robert James Mann

A Guide to the Knowledge of Life, Vegetable and Animal

ISBN/EAN: 9783337377045

Printed in Europe, USA, Canada, Australia, Japan

Cover: Foto ©berggeist007 / pixelio.de

More available books at **www.hansebooks.com**

A GUIDE

TO

THE KNOWLEDGE OF LIFE,

VEGETABLE AND ANIMAL;

BEING A COMPREHENSIVE

MANUAL OF PHYSIOLOGY.

VIEWED IN RELATION TO THE MAINTENANCE
OF HEALTH.

BY ROBERT JAMES MANN, M. D.

REVISED AND CORRECTED

NEW YORK:

C. S. FRANCIS & CO., 554 BROADWAY.

1860.

A GUIDE TO

THE KNOWLEDGE OF LIFE,

VEGETABLE AND ANIMAL

Being a comprehensive Manual of Physiology, viewed in relation to the Maintenance of Health. By ROBERT JAMES MANN, M. D. Revised and Corrected.

" This book is by one of the scientific teachers of the time; sound in knowledge, earnest in purpose, and above all writers on intricate subjects, gifted with wonderful powers of explanation and description. . . . . Whatever requires to be known of the portions of the body, their functions and uses, and the best means for their sustentation and healthy action, is here displayed, and intelligible at a glance. . . . . Nothing is omitted which can either gratify the curiosity or inform the mind."—*Chambers' Journal.*

TABLE OF CONTENTS.

PREFATORY REMARKS.

The first mariner who undertook to steer a ship out into the pathless ocean, did so in entire ignorance of the laws of the elements to which he was about to confide his charge, and of the position of the rocks and shoals that were likely to beset his course. His trust was altogether in fortune and accident. Now, however, in matters of seamanship, the case is greatly changed. Every individual who is destined for a responsible position as a navigator, is prepared for an efficient discharge of his duties by being carefully instructed in all that is known of the configuration of the earth, besides being taught to interpret the secrets of magnetism, the mysteries of winds and waves, the eccentricities of time, and the hieroglyphics of the stars. Vessels, laden with rich merchandise, are no longer placed in the hands of ignorance and incapacity, and left to be drifted to shipwreck or to their haven, as chance may determine. Intelligent and skilful pilotage is prepared for them beforehand, that they may be guided in safety amidst rocks and shoals, and that even the hurricane may be eluded or conquered should its fury be encountered by the way.

But men still do, with a far more valuable possession, what they have long ceased to allow with their merchandise. Thousands and millions, without scruple or hesitation, undertake to pilot their own bodies through the voyage of life, although they have not acquired one clear idea regarding the rocks and shoals they will of necessity have to pass amongst, or concerning the laws of the tempests they will no less surely have to brave. Human existence is a treacherous ocean, thickly studded with dangerous obstacles, and swept by the fiercest hurricanes. But the obstacles have for the most part fixed situations, which are accurately mapped down, and the hurricanes move in obedience to unalterable rules that have been established for their control. Science has shown that, in almost every instance, disease and premature death, occurring before the appointed term of threescore years and ten, are consequences either of ignorance or wilful perversity. They are, indeed, the shipwrecks which rashness and presumption encounter, but which intelligence and skill are able to avoid. The period is consequently at hand when a similar revolution must be brought about in physiological and social affairs, to that which has been long since effected in nautical matters. As population grows more dense, and as people crowd more and more into towns for the benefit of co-operation and association, an acquaintance with the main principles that are concerned in the preservation of healthy existence becomes absolutely a matter of life and death. Typhus fever and cholera register this fact at frequent intervals, in fearful characters that are intelligible to all. And the physiologist and physician read the same truth, day by day, inscribed in more recondite forms, in every spot to which they direct their steps. A strong conviction has, accordingly, been more and more manifesting itself amongst intelligent men, that no human creature ought ever again to be sent forth on the perilous ocean of social existence, with the responsible charge of his own wonderful organization upon his hands, without having had some clear sailing directions given to him; —some preliminary instruction regarding the constitution of his body, and the physical relations by which it is influenced. He ought at least to have charts and loadstones, chronometers and polar stars, as his familiar companions in the troubled voyage. Impelled by considerations of this nature, the most eminent physicians and medical philosophers of the metropolis have recently taken occasion to issue a deliberate declaration of their belief that "it would greatly tend to prevent sickness, and *to promote soundness in body and mind*, were the elements of physiology, in its application to the preservation of health, made a part of general education," and also of their conviction that "such instruction may be rendered most interesting to the young, and may be communicated to them with the utmost facility and propriety in ordinary schools." At the time when this declaration was made, the author, acting upon his own previous experience and observation, had just completed the preparation of the first edition of this little manual, with the aim of supplying the important need that was thus authoritatively proclaimed. In the favorable reception that has been accorded to the manual, he finds a pleasant assurance that his aim has not fallen wide of its mark.

In carrying out his plan of preparing a course of physiological instruction, that shall be adapted equally to the wants of schools and of the public at large, the

author has deemed it best to address himself *immediately* to the reason and intelligence of his readers.
He has endeavored first to teach the broad principles
upon which organization is based, and then to point out
inferentially how these broad principles apply to sani-
tary regulations and considerations. The advantage of
this proceeding over any more dogmatic handling of the
subject, is that the student becomes trained by it to
meet any new combinations of circumstances that may
occur in life, with a fair chance of seeing their bearing
correctly. He can apply broad principles in a thousand
different ways, as unforeseen occasions arise. But particular and definite directions are liable, in the ever-vary-
ing complications of social existence, to fail him at his
greatest need. It is obviously the better course that the
understanding should be possessed with the reason of
things, and should be then left to make its own practical
arrangements in accordance with its acquired insight,
rather than that it should be told merely that this or
that ought to be done. Accordingly the " Guide to
the Knowledge of Life" treats of vitality in the broad-
est and most philosophic sense. The chemical and physical laws that are concerned with the work of organization are first explained; the mutual relations and
compensations of vegetable and animal structure are
then indicated; and next, the material composition of
the several parts of the animated frame, and especially
of the muscular and nervous apparatus, is sketched.
After this, the constitution of the brain, and the connection of its substance with the faculties of instinct and
intelligence, are dwelt upon, mainly with a view of en-
forcing the great duties, and illustrating the capabilities

of a sound course of education. Incidentally to these
interesting topics, several considerations of the highest
practical moment are entered upon : such, for instance,
as the means by which the fresh air is made a hot-bed
of pestilence ; the course whereby food is turned into
poison, and drink into liquid venom ; and how sensual
indulgence saps and destroys the vigor both of body and
mind, whilst habits of rational self-control and refined
intelligence develope in both the highest and noblest
powers. Finally, the nature of disease, and the cause
and meaning of premature decay, are viewed in relation
to remedial and preventive measures. The "GUIDE TO
THE KNOWLEDGE OF LIFE" is, therefore, a comprehensive statement of the fundamental principles of physiological and hygieinal science.

But in order to fit them for educational use, the
principles set forth in this little manual have been
reduced to the form of compendious axioms. These
axioms will be found to have all the convenience, without any of the objections, of answers to direct interroga-
tion. Questions should be analytic, and should be employed only as tests of the comprehension of the learner.
When they are printed with answers, already prepared
to be given to them, they are not this. They then
merely favor laziness in the teacher, and incompleteness
of apprehension in the pupil, and the affair becomes entirely one of memory alone. The axioms of the manual
are so brief and clear that they may readily be grasped
and retained in the recollection ; but each one is accom-
panied by a running commentary of an illustrative character, printed in a slightly smaller type. This is intended
to be maturely deliberated upon in connection with the

main proposition, in order that its full force and meaning may be brought out and mastered. The axiomatic propositions are built up, one upon the other, in a sort of consequential relation, and they are arranged in numbered paragraphs, so that by the employment of numerical references they may be made to explain and illustrate one another, as occasion arises. When the manual is used in classes and schools, the teacher should find in these illustrative references and commentaries, abundant suggestions for testing the progress and proficiency of the pupil.

GUIDE

TO

THE KNOWLEDGE OF LIFE.

CHAPTER I.

ORGANIZED STRUCTURE.

1. A *structure* is any thing that is *built up* of subordinate parts.

A house is built up of bricks, and therefore a house is a structure.

2. An *organized* structure is any thing that is *built up of subordinate organs*, each having an office of its own to perform.

The term *organ* is derived from a Greek word (*organon*), which signifies an *instrument*. A plant is built up of cells, roots, branches, leaves, and flowers, all of which are distinct organs, or instruments, performing distinct offices ; consequently a plant is an organized structure. In the same way an animal is an organized structure, because it is built up of a stomach, a heart, blood-vessels, lungs, skin, and other separately-acting parts.

1   A

3. Organized structures *undergo continual change* in their *internal* composition ; but unorganized structures remain unchanged in their substance through long intervals of time.

Plants and animals take in food day after day, convert that food into sap and wood, or blood and flesh, and at the same time throw out parts of their old substance to make way for the new material. Stones, on the other hand, which are built up of unorganized parts, undergo no internal change of composition so long as they continue to be stones. In organized bodies, motion and change is the rule, and fixedness the exception, but in unorganized bodies the case is reversed ; in them change is the exception, and fixedness the rule.

4. *Life* is the *state* in which organized structures exist.

The physiologist employs the term life in a strictly *physical sense.* Plants and the lower animals possess life as well as man, and therefore no spiritual endowment is comprised within the motion thus expressed. Organized structures exist in a condition of continued activity and change, so far as the internal composition of their substance is concerned, from the instant of their organization, until the period when that organization is about to be finally dissolved. Hence it is found convenient to contrive a word that may be understood to express the active condition, so long as it lasts, even although the various facts, on which the condition depends, are imperfectly known. Life certainly is not merely a property of organized structure ; for bodies, that have been alive, retain traces of organized structure, even when they have lost their vitality. The word life is derived from the old Anglo-Saxon term (*lyfian*) for " to live," or " to remain."

5. The changes of internal composition, which living organized structures undergo,

depend upon two conditions. In the first place, the changing bodies must be *supplied with* suitable *food.* In the second place, they must be *surrounded by certain external influences,* capable of stimulating them to the performance of the change.

Both plants and animals die of starvation if food be withheld from them. But both also are frozen to death in the entire absence of heat.

To stimulate is to goad, or stir up; the word is derived from a Latin term (*stimulo*) that bears this signification.

A due supply of food is essential to the preservation of the life of organized structures, because that life is the state in which they exist, while their internal composition is *undergoing change.* There can be no change in the substance of bodies unless new matter is constantly furnished to take the place of the old as it is removed. Hence, if food be withheld from either plants or animals, all the activities of their living organs stop, and the organs themselves languish and die. Change of substance is indispensable to existence in the active state, called life. Vital actions are indeed as much connected with the change of matter into a new form, as the heat given out by the fire is connected with the change of the burning coal into smoke and invisible vapor. Life needs to be supported by food, as a fire requires to be kept up by fuel.

6. The *external influences,* that stimulate living organs to perform their appropriate work of change, and to convert food into their own substance, are *light, heat,* and *electricity.*

Every one knows how sluggish the vital actions of plants are during the gloomy and cold season of winter, and how brisk and vigorous they become with the return of summer. Animals are less dependent upon the external presence of light and heat than plants, but nevertheless require them in a modified extent. Neither animal nor vegetable life can be continued in absolute darkness and cold, however abundant the supply of nutritious food may be.

It is probable that electrical, and perhaps other even more subtile influences, are operative in causing living structures to undergo change of composition, as well as light and heat. Science has not, however, yet been able to furnish satisfactory proofs that such is the fact.

7. The *changes* of internal composition that are constantly going on in organized bodies, are of a *hidden nature*, and can rarely be observed by the eye.

Hence, it is not always possible to determine at once that a body is a living structure, by noticing the presence of vital operations. The only result of vital action, that the eye can directly perceive, is movement from place to place; but the power of effecting this is almost confined to the animal creation.

8. Organized bodies can, however, generally be at once *distinguished* from unorganized, by the *form* they present to the eye.

Organized bodies have generally regular forms that are bounded by curved and swelling outlines. Unorganized bodies, on the other hand, are either devoid of regular forms, or they have regular forms that are bounded by straight lines and angular corners. The only exceptions to this rule occur in a few of the lowest kinds of living creatures.

9. The several parts of unorganized structures are connected together simply by *proximity of position*.

The individuality in unorganized bodies is in the separate particles, or little parts, that conspire to form the whole. Numerous particles, which are all exactly like each other, are placed near together. This kind of gathering is called *aggregation*, from a Latin word (*aggrego—ad gregem ducere*), which means to assemble. Aggregation is the collecting together of a number of like particles into one mass.

10. The several parts of organized structures are connected together *by relations of mutual dependence.*

If all the roots of a plant are destroyed, the leaves and stems also die. No animal can go on living after its stomach, its heart, or its lungs have been taken away. Organization is the building up of unlike parts into *one body.* In organized structures, the individuality is in the body formed, and not in the forming parts.

11. Organized bodies always *stop in their growth* when they have attained a certain definite size. Aggregated bodies, on the other hand, may be of *any conceivable dimension.*

The whale, the length of whose body is bounded by some 80 feet, is the largest organized creature known. But rocky masses often form vast mountains. Nay, the stupendous earth itself is but one aggregated mass. Trees and corals sometimes acquire a very large size, but they are properly connected clusters of individuals, and not single bodies. They combine in themselves, so to speak, some of the attributes of both organization and aggregation, and thus form a sort of intermediate group.

12. Organized structures are generally of a *soft consistence*—whereas aggregated bodies are mostly very *hard.*

In unorganized bodies, hardness and solidity form the rule, and softness is the exception. But in organized bodies, hardness is the exception, and softness the rule. The softness results from an intermingling of solid and liquid matter, and it is one of the means taken for adapting the structure to the necessity of continued internal change of composition. Hardness is only given to organized bodies, when great capacity for resistance is required in them, as in the case of the trunks of trees, or of the bones of animals. Organized substances only become hard, like mineral substances, when they have similar mechanical offices to perform. Organs that are concerned in the immediate maintenance of life are always soft.

CHAPTER II.

THE ELEMENTARY MATERIALS OF ORGANIZATION.

13. The bodies of plants and animals are built up of organs, but those organs are, in their turn, formed of yet smaller parts, or, *particles*, which are therefore the rude materials of organization.

The organs of plants and animals are built up of particles, as houses are of bricks. "Particle" is derived from the Latin (*particula*) for a small part.

14. The *small particles* used in the construction of animal and vegetable structures, are called *ultimate atoms*.

The word *ultimate* is derived from a Latin term (*ultimus*), which signifies *last* or *utmost*. *Atom* is compounded of two Greek words (*a. temno*), that imply an incapability of *being cut or divided*. An ultimate atom is therefore a particle of matter that has been divided to the utmost; or, in other words, a particle so small that it cannot be further divided.

15. The *ultimate atoms*, of which organized structures are composed, are so minute that they *cannot be seen* when separated from each other.

The smallest visible grain of matter can be divided, by the processes of chemistry, into a vast number of yet more minute particles. The finest dust is a collection, not of ultimate atoms, but of little heaps of ultimate atoms. The ultimate atoms of matter, whether organized or unorganized, have never been seen by the human eye.

16. The *existence* of these invisible ultimate atoms, in material bodies, *is inferred* from certain mechanical and chemical effects that come under the notice of scientific men.

This statement will be found to be illustrated a few pages further on. Some philosophers have maintained that there can be no such things as ultimate atoms of matter, and that any given particle might be divided and subdivided for ever, without the end of its divisibility being reached, if we only knew how to set about the task. Common sense cannot comprehend the possibility of such endless subdivision. The consideration need not, however, be entered upon, as it is sufficient, for all practical purposes, to understand an ultimate atom as being a particle of matter which is so small that no employment of the human faculties can ever detect evidence of the possibility of further division.

17. The ultimate atoms of material masses *hang together by mutual attraction*. Each atom attracts all other atoms that are very near to it.

The bricks in a wall are held together by a cement called mortar. It will be easily understood that if the

bricks were made alternately of loadstone and steel, no
mortar might be required, as they would then stick
together by the influence of their own attractions. In
natural structures, no cements are used to hold together
the component atoms, but the atoms are endowed with
inherent attractive powers, which enable them to attach
themselves together with a considerable amount of force.

" Inherent" is derived from a Latin word (*inhæreo*),
which signifies "dwelling in."   An inherent property is
simply a property that dwells in any body in virtue of
its existence, and as an inseparable attribute.

18. The attraction, that is employed in
holding together the ultimate atoms of material bodies, is called the attraction of *cohesion.*

"Cohesion" is derived from a Latin word (*cohæreo*),
which signifies to stick together.   Cohesive attraction
comes into play when ultimate atoms are tolerably near
together.

19. The ultimate atoms of material bodies
are not *in close contact* with each other, even
when they adhere cohesively together to
form a connected mass.

All bodies can be compressed by the employment of
suitable degrees of force, and so made to occupy less
space.   When bodies are thus compressed, their component atoms must be squeezed more closely together.
It is highly probable that no amount of compression can
force ultimate atoms really to touch each other.   Each
atom, even in the densest body, seems to have a clear
space all round it, and, therefore, between it and its
nearest neighbors.   The ultimate atoms of bodies never
get into absolute contact with each other, because they
repel one another *very powerfully* when they get very
close together.   .

The power atoms possess of repelling each other, when
at close quarters, is conceived to be due to the presence

of heat. This inference is drawn because all bodies are observed to swell out into a larger bulk as they become hotter. When bodies grow larger, without the addition of any new material to their substance, it is clear their component atoms must be drawn further asunder. It is probable that the ultimate atoms of bodies would touch each other, if the masses they form could be entirely deprived of heat.

20. The *precise distances* at which the ultimate atoms of any given body remain from each other, are determined by the degree of heat that is present to become operative in *resisting the cohesive attractions.*

The more heat, or repulsive force, any substance contains in an active state, the further its ultimate atoms rest from each other, and the larger is the bulk it fills. All bodies expand their dimensions when heated, and contract them when cooled. An iron rod, that just fits into a metal ring when cold, cannot be forced within it when made red hot.

21. Bodies may be heated until their ultimate atoms are driven *so far asunder* that they cease to be found cohesively together.

The ultimate atoms of all kinds of bodies are loosened from their cohesive union, if a sufficient amount of heat be applied. When a lump of lead is placed in an iron ladle set over a fierce fire, the cohesion of its ultimate atoms is very soon suspended, and the atoms run about freely amongst each other. The same thing happens with silver, gold, stones, and indeed all solid bodies, of whatever kind, if a sufficiently elevated temperature be employed.

22. When the cohesive attractions of the ultimate atoms of a body are so far weakened, that the atoms can move freely about with

A*

regard to each other, the body is said to be *liquefied.*

The ultimate atoms of *solids* are very close together, and attract each other so very powerfully, that they cannot move their positions with regard to each other, unless driven to do so by the application of a great degree of force. The ultimate atoms of *liquids* are further asunder, and attract each other so slightly, that the application of a very trifling impulse is sufficient to make them change their relative positions. If the finger be pushed against a piece of lead, the atoms of the lead hold together so firmly, that the finger cannot get in between them; either all the atoms of the lead move together, or none of them stir. But if an iron rod be pushed against the surface of a mass of molten lead, the rod sinks into the mass, or, in other words, gets in between its particles, which move asunder to make way for the iron. "Solids" is derived from a Latin word (*solidus*), that signifies "the whole" (connected into a whole)—"liquid" from a Latin word (*liqueo*), that means to melt.

23. But the ultimate atoms of a body may be driven *further asunder*, than they are when that body is in the liquid state.

The ultimate atoms of all liquids get considerably further from each other, when very high degrees of heat are applied. If a piece of ice be placed before a hot fire, its atoms get separated to a certain extent, and the ice is liquefied into water. If the water be then boiled over the fire, the atoms fly still further asunder, and the water becomes steam. When a cubic inch of water is converted into steam. it is made to occupy seventeen hundred times more space than it did before the change; consequently, its several particles must be further asunder than they were before.

24. When the ultimate atoms of a body are placed *so far asunder* that they cease to exert any effectual attractive influence upon

one another, the body is said to be in a *gaseous state*.

In steam, the ultimate atoms fly quite away from each other, unless they are confined or constrained by some external influence—the power of mutual repulsion, conferred by heat, continues to act at very much greater distances than cohesive attraction can. Hence, long after atoms have got so far asunder that they cease to cling together by mutual attraction, they go on spreading themselves still more widely apart. · The atoms of solids *confine each other* by the strength of their mutual attraction; hence a lump of any solid substance preserves its form when laid upon a table. The atoms of liquids *cling together* without confining each other; hence, water runs along in a stream when poured upon a table; liquids must be placed in glasses and jugs, if they are not to be allowed to flow away under the simple operation of their weight. The atoms of gases *actively repel* each other, without exercising any confining or clinging influence of a counteracting kind; hence, when a solid or liquid is converted into a gas, it falls asunder, and is scattered into air. Gases must be placed in closed bottles, or confined vessels, if they are not to be allowed to escape, for their atoms are able *to flow upwards* out of open vessels, under the influence of their mutual repulsions, which more than antagonize the power of weight. "Gas" is derived from a German word (*gheist*), which signifies spirit.

25. *Gaseous bodies* are generally *invisible* and *transparent*, instead of being visible, as solids and liquids are.

Solids and fluids are visible, because their ultimate atoms are so near together that several can be seen as one. When a body, whose ultimate atoms are too small to be visible, falls to pieces, it must become altogether invisible, if every atom separates from its neighbor, and all get so far asunder that no two can be seen together as one. It can be shown to be highly probable, that the ultimate atoms of gases are at least 100 times their own diameters asunder, even when those gases are held in confined vessels. Transparency is often due to other causes, but there can be no doubt that a body which is invisible, in consequence of the minuteness of its atoms, and of the distance at which these are placed, must also be open to the passage of the undulations of light.

There are some gaseous bodies that have a slight tinge of color, and that, therefore, can be seen. These, however, are exceptional cases. Nearly all gases are quite invisible, like high pressure steam when it bursts from the safety-valve of a locomotive, or like air.

26. *Different bodies* require *different degrees of heat* to effect their conversion into the gaseous state.

Every known body can be converted into the gaseous state, if a sufficiently high degree of heat be employed. But different bodies require different degrees of heat to effect the result. The fixed air which escapes when bottles of effervescing wine, or soda water, are opened, remains in a gaseous state, until placed in a temperature as much colder than ice as boiling water is hotter than ice. Water requires 212 degrees of heat (reckoned by Fahrenheit's scale) to convert it into a vapor. Mercury is not changed into vapor until heated to 660 degrees.

27. Bodies, that can only be retained in the gaseous state while at high temperatures, are then said to be *vapors*.

Thus the fumes of mercury are properly a vapor rather than a gas. "Vapor" is derived from the Latin word (*vapor*) for "steam."

28. Bodies that remain in the gaseous state at ordinary temperatures, are called *gases*.

Hence, the fixed air, that escapes from effervescing soda water, is a gas, and not a vapor. A gas is properly the vapor of a body that boils at a lower temperature than that which is present naturally at the earth's surface. Hence, gases are never seen in the solid, or liquid state, unless subjected to artificial influences.

29. All that portion of the earth's substance that remains in the gaseous state at ordinary temperatures, is placed round the denser solid part as an *atmosphere*.

The great bulk of the earth's gaseous material is piled up around and above the organized creatures that dwell upon its surface; and is, indeed, the air that animals breathe. The word atmosphere is derived from two Greek terms (*atmos-sphaira*,) that signify a sphere or globe of vapor. "Air" is taken from a Greek word (*ao*), that means "to breathe," or "blow."

30. The atmosphere, or gaseous investment that surrounds the solid surface of the earth, is a *reservoir of the materials* employed in organization, kept ready for present and immediate use.

The ultimate atoms of organized bodies are originally procured from the unorganized masses of the earth. Even the flesh and blood of man are taken from the dust of the ground. There is no ultimate, elementary atom, in the structures of plants or animals, that the chemist does not also find to be present in the structureless, aggregated masses of the globe. But, so long as these atoms are held together in the state of dense solids, they are not in a convenient condition for the purposes of organization. Hence nature keeps a store, of such as are of most immediate importance in the work, loosened out and ready for service. It is this store of loosened-out atoms, prepared for the work of organization, that constitutes the atmosphere.

CHAPTER III.

COMPOSITION OF THE ATMOSPHERE.

31. THE invisible atmosphere is known to be really a material substance, because it *occupies space* to the exclusion of other bodies; because it *possesses weight;* and because it *pushes* heavy bodies before it, when in motion.

If a glass tumbler be plunged mouth downwards into water, the water is pushed away before the open mouth, just as it would be, if the glass were filled with solid substance. The inverted glass is full of air, and therefore cannot receive water, until the air is removed.

A glass flask weighs more when it is full of air, then it does after the air has been pumped out; hence, it is clear that air itself possesses weight. The column of mercury ascends nearly thirty inches into the empty tube of the barometer, because the weight of the atmosphere, acting upon the surface of the reservoir of mercury outside the tube, forces it up so far. A column of air of the same transverse dimensions as the inside of the barometer tube, but extending to the upper termination of the atmosphere, more than 80 miles high, weighs as much, then, as a column of mercury of the same breadth, and about 30 inches high. Air is 14.000 times lighter than mercury. One hundred cubic inches of air (near the ground, and at ordinary temperatures) weigh **31** grains.   100 cubic inches of mercury weigh 313,216 grains.

Wind (in other words, air in motion) causes mill-sails to whirl round before its pressure, and drives laden ships across the sea.   The force, by which these mechanical effects are produced, is strictly analogous to the power that enables the cannon ball to plough its fearful way through opposing obstacles.   All moving bodies are able to impress other bodies, at rest, with their own motion, upon striking them ; but then they lose exactly as much moving force as the bodies, that are struck, receive.   Mill-sails and laden ships receive motion when struck by the wind ; hence, the wind is a moving *body*.

32. The material atmosphere is made up of *more kinds* of gaseous substance *than one.*

Since the ultimate atoms of gaseous bodies are many times their own diameters asunder, it follows, that several different gases may be easily diffused together through the same space, the atoms of the one kind occupying the intervals that lie between the atoms of the other.   Different gases are mingled together, in the atmosphere, in this way.

33. The atmosphere contains within itself a gas, that is capable of *sustaining flame* and animal *life.*   But it also comprises within itself a gas, that *extinguishes* both *flame* and animal *life.*

If water be poured into a soup plate, a piece of wood or cork, with a fragment of wax taper fixed on it, be floated on the water, the taper be lit, and a tall glass jar be then placed over it, so that it rests in the water, it will be found that the flame consumes a portion of the air, as the taper burns, and that the water rises within the jar to take the place of the substance consumed.   After a time, however, the flame of the taper is extinguished, because there no longer remains within the jar any thing capable of sustaining it.   There were, therefore, two different kinds of gases, at first, in the jar; one that was able to sustain the burning of the taper, and that was itself absorbed, or removed in doing so; and another that was quite unable to sustain the burning, and that, therefore, remained unconsumed and uninfluenced, while its companion was engaged about the work.

When a bird or a mouse is confined within a closed jar, much the same result happens. The animal continues to live, until a portion of the air has been consumed by its breathing; but, after that has happened, the animal dies of suffocation.

34. The flame and life-sustaining gas, contained within atmospheric air, is called *oxygen.*

The word "oxygen" is derived from two Greek terms (*oxus gennao*), that signify "to produce sourness." The name was adopted because the early chemists found that a great number of fluids became sour when they imbibed oxygen. Lavoisier thought that there was no other cause of acidity than the presence of this gas.

35. Flame-sustaining and life-sustaining *oxygen is marked out* from all other *elementary* bodies in nature, by its possessing a very wide range of energetic affinities.

All such bodies as have resisted the attempts of science to reduce them into simpler forms of existence, are called *elements.* The word is taken from the Latin term for a rudiment or first principle (*elementum*). Oxygen may be characterized, for distinction's sake, as the *energetic element* of material nature. Its activity is incessant. It leaves scarcely any thing free from its modifying influence. It rusts our iron—consumes the fuel of our fire while it burns—turns fermented liquors sour—corrodes nearly every solid substance it comes into contact with, and wastes even the membranes and fibres of living frames. "Energetic" is derived from a Greek word (*energeia*) which signifies "activity."

36. Activities, such as oxygen thus energetically displays, are termed, in chemical language, *affinities*.

As ultimate atoms of a *like* kind are bound together by the attraction of *cohesion*, so ultimate atoms of an *unlike* kind are bound together by a different form of attraction, which is then termed chemical attraction, or the attraction of *affinity*. The word *affinity* is derived from a Latin term (*affinis*), which signifies "near relation," or more strictly, "relation by marriage." The union of unlike atoms of matter is thus likened to a matrimonial alliance. The attraction of affinity is sometimes spoken of as *chemical* attraction, because it is the one attribute of matter with which chemists have especial concern.

37. When unlike atoms are bound together by chemical affinity, *compound atoms* are formed by the union.

"Compound" is derived from a Latin word (*compositus*), that signifies "laid together." The ultimate atoms of elementary bodies are bound together cohesively to form masses; but, in compound bodies, unlike ultimate atoms, are first united together by the attraction of affinity to constitute compound atoms, and these compound atoms are then cemented cohesively into masses. Thus iron is a simple substance; a lump of iron is merely an aggregation of ultimate atoms that all resemble each other. Iron-rust, on the other hand, is a compound body; in it atoms of rust are joined together by aggregation, but in each atom of rust there are atoms of iron and oxygen united by affinity.

38. Oxygen possesses so *wide a range* of affinities, that it is found combined with nearly all the other kinds of elementary atoms that are known.

Sixty-four different kinds of elementary bodies are

now known. The ultimate atoms of oxygen contract
unions of affinity with the ultimate atoms of nearly all
these bodies. By means of these unions, the great bulk
of the earth's superficial substance is formed. About
half the weight of the material substance that has come
under the notice of man, is made up by oxygen alone.

39. Most of the elementary bodies combine
with oxygen, *in preference* to contracting any
other kinds of chemical union among themselves.

The chemical attractions of oxygen are as energetic as
their range is wide. Some kinds of unlike atoms are
more prone to contract alliances of affinity than others.
But oxygen dissolves most other alliances that have been
formed without its presence; hence, its attractions of
affinity are known to be stronger than the attraction
exercised by any other kind of body. It is through the
instrumentality of this sort of *preference for certain
kinds* of union, among ultimate atoms, that all the great
chemical operations of nature are carried on. Whenever simple atoms, possessed of strong attractions, are
brought near to compound atoms, whose elements are
less powerfully held together, the union, that had been
previously formed, is dissolved, and a new one is arranged.

40. The *sensible characters* of combining
atoms are altogether *changed* when chemical
union takes place.

A case of chemical union can, in this way, be at once
distinguished from a case of cohesive aggregation. When
atoms of iron are united cohesively with atoms of iron,
the mass formed still possesses no other character than
that of iron. But when atoms of oxygen unite chemically with atoms of iron, the resulting mass ceases to
have either the properties of oxygen or of metallic iron;
it is neither a transparent gas like the one, nor is it a
metallic substance, attracted by the magnet, like the

other. It is then the dull red powdery substance known as rust.

41. *Compounds* that contain oxygen, therefore, present *none of its characters*, so long as the state of chemical union continues.

The rust of iron, and the powdery substance known as the black oxide of manganese, both contain oxygen, but none of the characters of the gas can be discovered in either of these substances, until its chemical union with the other elements there present is dissolved.

42. Oxygen may be procured in a pure and *simple state*, by driving it off from some of the compound bodies that contain it, through the influence of heat.

If the touch-hole of an old gun-barrel be plugged up, some black oxide of manganese be placed in the gun, and a piece of flexible tube be attached to the mouth of the barrel, pure oxygen escapes from the tube when the breech is made red hot in the fire. The escaping gas may be caught in a jar, inverted over, and filled with water, by placing the end of the tube beneath the rim. The gas will then bubble up into the glass and slowly displace the water. In this way, oxygen may be procured for the purpose of examination.

43. Oxygen gas is a *light, invisible,* and *transparent* substance, like the air which it helps to form.

Oxygen is a trifle heavier than an equal bulk of air. One hundred cubic inches weigh 34 grains, near the earth's surface, and at ordinary temperatures (31).

44. Oxygen gas is devoid of *taste*, or *smell*, and, to a great extent, resists the solvent power of water.

The presence of oxygen cannot be *directly* detected by the senses, any more than the presence of impalpable air. Its existence has been only ascertained through certain manifest effects that it produces. Water dissolves a very little of the gas when placed in contact with it.

45. Oxygen produces *acid compounds*, when it unites chemically with some of the other material elements.

When phosphorus is burned in a jar of oxygen, white fumes are formed, which consist of a combination of phosphorus and oxygen. These fumes are readily absorbed by water, and they then communicate a sour taste to the liquid. There are 11 elementary bodies with which oxygen unites to form acid compounds.

46. Oxygen produces *compounds which are not acid* when it unites chemically with others of the material elements.

Oxygen forms compounds which are not acid when combined with one or other of 51 distinct elements. These non-acidulous compounds of oxygen are classed together by chemists as oxides, to distinguish them from the acid compounds. It is a remarkable fact that the acidulous oxygen-compounds again unite with the non-acidulous oxygen-compounds, to form threefold or ternary combinations, which are no longer acid. These ternary, neutral combinations are called salts. Thus phosphorus and oxygen unite to form an acid, which is called phosphoric acid by the chemists. Oxygen and iron unite to form iron-rust, which is not acid. Phosphoric acid and iron-rust (or oxide of iron) unite to form a compound which is no longer acid, but possesses the properties of a salt, and is called phosphate of iron. This may be expressed to the eye in the following way.

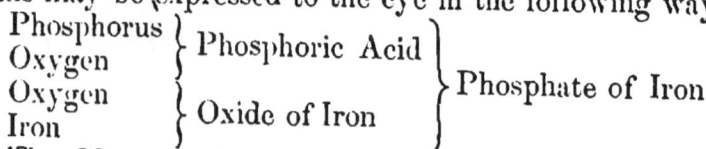

$$\left.\begin{array}{l}\left.\begin{array}{l}\text{Phosphorus}\\\text{Oxygen}\end{array}\right\}\text{Phosphoric Acid}\\\left.\begin{array}{l}\text{Oxygen}\\\text{Iron}\end{array}\right\}\text{Oxide of Iron}\end{array}\right\}\text{Phosphate of Iron}$$

The fifty-one elementary bodies, with which oxygen

unites to form the non-acidulous compounds, are all of the nature of metals. The other eleven, with which it forms acid compounds, are therefore distinguished as being the non-metallic elements. Oxygen thus seems to stand between two opposite classes of material elements, and the whole may be classified thus:

11 Non-Metallic Bodies
Oxygen
51 Metals

There is a 64th element, called fluorine, with which oxygen does not unite, and which often seems able to play the part of oxygen, to a certain extent, in combining itself with other elements.

47. Whenever oxygen combines very rapidly with other elementary bodies, *light and heat* are evolved.

All the ordinary forms of combustion are illustrations of the proneness of oxygen to enter into combination with other elementary bodies. The burning of combustible substances is merely the conversion of their ultimate atoms into compound atoms of a gaseous nature, by union with atoms of oxygen. The waste is apparent and not real; it is simply the change of a visible substance into invisible compounds. The light and heat, that constitute the visible characters of the combustion, are set free as a consequence of the combination. The peculiar, chemical energy of oxygen is strikingly exhibited when highly heated bodies are plunged into jars containing only the pure gas. Combination then goes on so rapidly that a most brilliant display, either of flame, or showers of sparks, results. Even iron-wire burns readily if introduced, when red hot, into oxygen.

48. A white triangle, thus formed △, may be conveniently used *as a symbol*, to represent to the eye an ultimate atom of oxygen.

Nothing whatever is known regarding the form of the ultimate atoms of matter; but it is assumed as highly

probable that different kinds of atoms possess different
forms. It is convenient to adopt a definite figure as a
symbol of an ultimate atom, and to use different figures
to symbolize different atoms, because this plan enables
the nature of certain chemical changes, that will have to
be described, to be expressed, at once, pictorially to the
eye. It must, however, be carefully borne in mind that
the figure is employed merely as a symbol, and does
not really represent any ascertained fact regarding the
form of the atom represented.

"A symbol" is a sign put to represent something.
The word is taken from the Greek (*sumbolon*), for a
sign.

49. The *life and flame-extinguishing* portion of the atmosphere is called *nitrogen*.

The word is derived from two Greek terms (*nitron
gennao*), which signify the " producer of nitre." Nitrogen
is so called, because the early chemists found that it was
a constant ingredient in nitre. Lavoisier discovered the
presence of this peculiar substance in atmospheric air,
but he gave it the very expressive name of *azote*, a word
derived from Greek terms (*a. zao*) that imply the " de-
privation of life."

50. Nitrogen is marked out from all other
kinds of elementary bodies by the *weakness
of its affinities*.

Nitrogen seems to be as backward in entering into
combination with other elementary bodies, as oxygen is
forward in doing so, and whenever it is trapped into contracting a chemical union, it always manifests a marked
tendency to escape from the position of constraint, as
soon as it can. As oxygen has been termed the *energetic
element* of material nature, nitrogen may be distinguished
as the *inert element*. " Inert" is derived from the Latin
word (*iners*) for " lazy."

51. Nitrogen is found in very *few* of the
unorganized compounds of the earth.

Nitre, and three other analogous salts, that occur in nature in small quanities, contain nitrogen. But these are the only unorganized compounds, formed independently of man's agency, in which it is present.

52. The *great bulk* of the nitrogen of the earth exists in a *free state* in the atmosphere. This is exactly what might be expected where so inert a body is concerned. The great bulk of the element continues in an uncombined state, and only a very small proportion of its atoms enter at any one time into a state of combination with other bodies. The great bulk of energetic oxygen, on the other hand, is kept fast locked in union with the earth's solids, and only a very small proportional part is left floating free amidst the nitrogen.

53. Nitrogen is left behind, in its separate and uncombined state, whenever *all the oxygen is abstracted* from a given quantity of atmospheric air. If a tall glass jar be inverted over a piece of burning phosphorus, placed on a small metal stand and surrounded by water, instead of over the burning taper, as described in a preceding paragraph (33), portions of the burning phosphorus unite with the oxygen comprised in the air, and the two form a white fume, which is then slowly absorbed by the water, so that at last nothing remains behind in the upper part of the jar but pure nitrogen.

54. Nitrogen gas is a light, *invisible*, and *transparent* substance, like the air which it helps to form, and is also perfectly devoid of *taste* and *smell*. Nitrogen gas has about the same weight, bulk for bulk, as atmospheric air. The senses do not enable an observer to distinguish directly between the simple element and the compound body.

55. But nitrogen gas is *entirely devoid of the corrosive properties* of common air.

Bodies, that soon get corroded and injured by exposure to air, remain perfectly safe and uninfluenced, when immersed in pure nitrogen. Combustibles, that are already burning, burn no longer when introduced into it. Flame goes out when plunged into it, exactly as if it were plunged into water. Living animals cease to breathe when confined in it. By this inertness, then, nitrogen may be at once distinguished from either oxygen or atmospheric air.

56. A black triangle, thus formed ▲, may be conveniently used, *as a symbol*, to represent to the eye an ultimate atom of nitrogen.

It must, however, be carefully remembered, as in the case of oxygen (48), that this figure is merely an artificial symbol, adopted for the purpose of convenient illustration.

57. Oxygen and nitrogen are mingled together in atmospheric air, in about the proportions of *one part* of the former, to *four parts* of the latter.

When phosphorus is burnt within a glass jar, inverted over water (53), the water rises, as the oxygen is removed by the burning body, until a fifth part of the jar is occupied by it. The exact proportions of oxygen and nitrogen contained in any quantity of atmospheric air, are 21 parts of oxygen to 79 parts of nitrogen (estimated by volume), or 8 parts of oxygen to 28 parts of nitrogen, if the estimate is made by weight; 36 grains of atmospheric air contain 8 grains of oxygen, and 28 grains of nitrogen.

58. The oxygen and nitrogen, in the air, are not *chemically combined* together; they are merely mechanically mixed, the atoms of the one amidst the atoms of the other.

This is evidenced by the fact, that all the special properties of oxygen are present in the air. The atmosphere corrodes bodies, and sustains flame and animal life, just as pure oxygen would do. The properties of the oxygen are in no way altered by the presence of its inert companion, excepting that its activities are slightly hampered and dulled by the association. It will be remembered that when elementary bodies enter into chemical union with other bodies, they lose their distinguishing characters, the compound acquiring certain specific properties of its own (40). Oxygen and nitrogen, combined chemically, form the highly corrosive substance known as nitric acid, or aqua fortis. Whenever unlike gases are so placed that they can get together, they invariably mingle mechanically, until they are equally mixed. If a bottle of oxygen and a bottle of nitrogen were deposited side by side, and the interiors of the two were then connected by means of a small bent tube, it would be found, after a few hours, that each bottle contained an equal proportion of oxygen and nitrogen intrinsically mingled together. The ultimate atoms of each kind of gas repel one another, without being in any way influenced by the presence of the atoms of the other kind of gas. One gas spreads through another, exactly as it would spread into empty space. This tendency of gases to intermingle freely with each other is technically known as their *diffusive power*. The term "diffuse" is taken from the Latin word (*diffundo*), for "to pour out," or "spread abroad."

59. The atmosphere consists, then, of the *chemically active gas, oxygen,* diffused amidst a large proportional quantity of *inert nitrogen.*

It has been seen that a large amount of free nitrogen might be expected to float around the dense solid nucleus of the earth as a sort of atmosphere, if only on account of the inertness of its chemical character. But if, at any time, free oxygen were thrown into this atmosphere of nitrogen, the two would become ultimately mingled together, under the influence of the law of gaseous diffusion, and, on account of the inertness of the nitrogen, they would in no way interfere with each other, unless under especial and exceptional circumstances. (Small quantities of oxygen and nitrogen combine chemically together during the electrical discharges of thunderstorms.) The two would exist in tacit harmony, each retaining its own attributes and character, notwithstanding the close association. The more active member of the alliance would merely be rendered a trifle more slow in its operations, in consequence of the listless neighbor that was always standing in its way—so to speak. It will be seen hereafter, that a provision is made for keeping up the due proportion of oxygen in the atmosphere, notwithstanding the large amount that is constantly being withdrawn from the free condition, under the influence of its own corrosive operations. The atmosphere is thus *oxygen diluted* and sobered down *by diffusion in the midst of nitrogen*, as vinegar is diluted when it is diffused in water. The *perfect inertness* of nitrogen renders it a most admirable agent for this office of dilution. Nitrogen is so inert, that when it suffocates animals immersed in it, it does so merely by depriving them of their supply of oxygen, and not by any positively noxious influence of its own. If it were itself positively noxious, it would still do harm while oxygen was mingled with it.

60. The free gases that form the atmosphere are piled upon the earth's surface to a *considerable height.*

Air is 11,000 times lighter than an equal bulk of mercury; consequently, the atmosphere *must* go up 11,000 times higher than the 30 inches, at which it sustains the column of mercury, in the tube of a barometer—that is, it *must* extend 11,000 times 30 inches, or 27,500 feet (a little more than 5 miles). In reality it stretches far beyond 5 miles. When Du Luc mounted in his balloon to a height of 20,000 feet, he still found that there was air enough above him to sustain the column of mercury in the barometer tube at 12 inches. It is probable that some atmosphere still exists at a height of 80 miles, although in so thin a state as to be there inappreciable to the human senses.

61. The gaseous substance of the atmosphere is *compressible*.

That is to say, the component atoms of its substance get squeezed closer together when compressing force is applied, and the diminution of bulk is always in exact proportion to the pressure. A cubic inch of air is reduced to half a cubic inch, when compressed by a doubled force, and to a quarter of a cubic inch when compressed by a quadrupled force. Hence, the air near the earth's surface is much denser than it is in the higher regions, for it is there squeezed down by all the pile of aerial substance above it. Every square inch of air, near the level of the sea, bears a column of higher air, amounting in all to as much as weighs 15 pounds, and its density at that level is the result of the pressure of these superincumbent 15 pounds.

62. The gaseous substance of the air is very *elastic*.

Air expands, as it is relieved from compression, and this expansion goes on to a seemingly unlimited extent, as the pressure is removed. When the compressing force is halved, the bulk of the air becomes doubled; when the compression is diminished to one-fourth, the bulk of the air is quadrupled. If a cubic inch of air be carried up from the level of the sea, to a height slightly exceeding 3 miles, it is taken above one-half of the atmospheric substance, and is consequently pressed upon by only 7 pounds' weight and a half, instead of 15, and becomes there two cubic inches instead of one. If the cubic inch be carried still higher, it expands to a still larger bulk. This is why it is that the air extends more than 27,000 feet away from the earth's surface. If the air did not get thinner in the higher regions of the atmosphere, a column of it 5 miles (27,500 feet exactly)

high, and one inch square, would weigh 15 pounds ; but
as it does get thinner in the higher regions, it takes a
column considerably more than 5 miles high to make up
the 15 pounds.

63. The aerial substance is *kept together* by
the influence of the earth's attraction for its
atoms.

As the atoms of the gaseous atmosphere are not held
together by cohesive attraction, and as they repel each
other under the influence of heat, they would get scattered asunder, through space, if no controlling influence
were at work to prevent this result.   The aerial atoms
do not thus wander off into space, because each is attracted towards the earth's mass, by a force that is
amply sufficient to prevent its doing so, although not
sufficient to prevent each rising up from its repelling
neighbor to a certain distance.   The atoms of the air
float at a given distance from each other, kept so far
apart by the power of repulsion, and prevented from
getting further by the weight by which each presses
upon its lowest neighbor's repulsion.   The atoms of a
solid rest at a given distance from each other, under the
opposing influence of repulsion and cohesive attraction,
and the earth's attraction then acts upon the connected
mass as a whole.   The weight of the air is the result of
the earth's attraction for its atoms.   The air is kept
upon the earth's surface by its weight, as much as a
gold coin is kept upon a table, when laid there, by its
weight.   But as the air is a gaseous and compressible
substance, its atoms are further asunder from each other
in the higher regions than in the lower, because in the
latter, its atoms are compressed together by the weight
of the higher aerial substances resting upon them, as
well as by their own tendency towards the earth.

64. The atmosphere is *bounded externally*
by a definite surface.

The atmospheric substance spreads itself out more
and more thinly, under the influence of its elastic repul-

sions and of diminished pressure, until, at last, the farthest atoms get into a position where the upward tendency of repulsion is no longer able to raise the downward-tending atom further, against the force of the earth's attraction. The power of repulsion is so far weakened, on account of the wide separation of the repelling atoms. This position marks the limit of the atmosphere. If an aerial atom were lifted beyond this limit, it would fall back to the *surface of the atmosphere*, when again left to itself, exactly as water drops back to the surface of the liquid ocean when left so to do. The exact position occupied by the surface of the aerial ocean has not been yet ascertained, but it is probably somewhere about a hundred miles away from the solid matter of the earth.

65. Gaseous air transmits pressure equally *in all directions.*

If a bladder of air be laid upon a table, with one of its sides resting against a heavy box, and the hand be then placed firmly on its opposite side, any downward pressure made on the top of the bladder will be felt to be communicated through the air laterally to the hand.

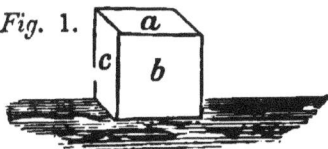

Let *a* in the figure (*Fig.* 1) represent a cubic inch of air resting on the sea shore, then the top *a* supports a column of air of 15 pounds' weight. But, in order that it may do so, *each* of the sides (*b c*, and those opposite) must also be sustained by a pressure equal to 15 pounds. If the pressure on the side *c* were only equal to fourteen pounds, then that resistance would have to give way before the fifteen-pound pressure, exerted at *a*, and the air would *flow* in a stream towards *c*. All the movements of the wind are due to unequally distributed pressures of this kind. In consequence of the great elasticity of the atmospheric substance, disturbances in the equilibrium of its atoms are very easily produced. The surface of the aerial ocean is constantly agitated by vast waves running miles high. The variations in the height of the mercurial column of the barometer indicate the movements of some of the larger of these atmospheric waves.

66. The atmospheric substance can be *perceived by the eye*, when the glance is directed up through its thickness, and towards void space.

The blue sky is the air seen bathed in faint earthshine, reflected up during daylight from the terrestrial surface. The air is not perfectly penetrable by this faint light, although it is so by the stronger light of direct sunshine; consequently the faint light is arrested amongst the aerial particles, and returned to the eye of the observer as blueness.

----◆----

CHAPTER IV.

WATER.

----

67. WATER *contains* in itself corrosive *oxygen.*

A piece of iron gets rusty when placed in water. The rust is composed of the pure metal combined with oxygen (45). The oxygen is taken out of the water to be combined with the metal.

68. Water contains in itself *another gaseous body* also, that is quite distinct from the oxygen.

If a quantity of iron filings are placed in a bottle

with some water and sulphuric acid, effervescence takes place, and a light gas escapes in a stream from the mouth of the bottle. This gas also comes out of the water. The iron filings take oxygen from the water, and are turned by it into rust, (which is then dissolved by the sulphuric acid and water,) and the substance, from which the oxygen is separated, escapes free from the mouth of the bottle.

69. The substance, which is combined with oxygen in water, is a *light inflammable gas.*

If a candle be held to the gas, as it pours out from the mouth of the bottle containing the iron filings and acid, it immediately ignites and burns with a pale blue flame.

70. The inflammable constituent of water is an elementary substance, and is called *hydrogen.*

The word hydrogen means "generator of water," (from the Greek, *udos gennao*). If a cold empty glass is held over the flame of burning hydrogen, the glass soon gets covered over with a thick mist inside. This mist condenses more and more, until at last it trickles down in drops, and it is then found to be pure water. The burning hydrogen unites with the oxygen surrounding it in the air, and forms water by the union.

71. Water, then, consists of *oxygen* gas and *hydrogen* gas, chemically combined together (37).

If oxygen and hydrogen gases, in the proportion of one part (by bulk) of the former to two parts of the latter, be placed together, in a glass tube, and an electric spark be passed through the mixture, an explosion takes place, the gases disappear, and a drop or two of water is found in their place. It is known that the gases unite *chemically* together, because the compound formed possesses new characters of its own, instead of having the characteristics of either of its elements (58).

In 100 grains of water, 11 grains are hydrogen, and 89 grains oxygen. There is twice as much hydrogen as oxygen, if the estimate is made by bulk; but this is because the hydrogen is eight times as light again as oxygen.

72. Hydrogen gas is a *light, invisible, air-like* substance, devoid of taste or smell.

Hydrogen gas cannot be distinguished from oxygen, nitrogen, or atmospheric air, by the direct observation of the senses. When a bottle is filled with it, it looks only like a bottle of air (43–54).

73. Hydrogen is distinguished from all other gaseous bodies by its *extreme lightness*, and by its *inflammability*.

Hydrogen gas extinguishes animal life and flame placed in it, like nitrogen (55); but it burns when a high temperature and a due supply of oxygen are afforded. The gas used in illuminating the streets and shops of towns is hydrogen, combined with another principle, to increase its light-giving power. Hydrogen is also so much lighter than nitrogen, that it may be poured upward into inverted jars, through air. Hydrogen gas is used to inflate balloons, on account of its lightness. The greater force with which the heavier ai, struggles to the earth displaces the ball of light hydrogen, so that this is forced to mount upwards in the air Hydrogen gas is the lightest body known. 100 cubi. inches weigh less than four grains (31).

74. A dotted triangle, thus formed △. may be conveniently used, *as a symbol*, tc represent an ultimate atom of hydrogen.

As in the case of oxygen, this figure must be taken only as an artificial symbol, adopted for convenience sake (48). It will be observed that the atoms of the three gases, oxygen, nitrogen, and hydrogen, are all represented by triangular symbols. These gases are all elementary bodies that remain in the gaseous state

at ordinary temperatures. Chemists have entirely
failed in their attempts to reduce them into yet simpler
states of existence.

75. A white and a dotted triangle, thus
conjoined ◁▷, form a symbol that conveniently represents a distinct *atom of water.*
The atom of water is a compound atom. It is composed of an ultimate atom of oxygen, and an ultimate
atom of hydrogen (37).

76. Water is a bland, *clear liquid,* at ordinary temperatures.
Water, when pure, is nearly transparent, and alike
devoid of color, taste and smell.

77. Water is *neutral in its chemical relations,* uniting alike with acids and oxides
(45), and dissolving most bodies it is placed
in connection with, without, however, producing any change in their sensible characters.
Solution is a kind of combination, standing midway
between chemical union (40) and mechanical admixture
(58). When salt or sugar is dissolved in water, the
atoms of these bodies and of the water are so intimately
mixed up together that they cannot be separated again,
otherwise than by driving off the water by heat. The
distinctive characters of the salt or sugar are, nevertheless, preserved unchanged in the solution. It tastes salt
or sweet; and the salt or sugar is found to be entirely
unaltered when the water is driven away from it. This
power of water to unite itself closely with numerous
bodies, without effecting any change in their essential
characters, is of great importance in numerous points
of view.

78. Water passes into the *solid state* in
the greatest cold of our winters (22).

Water is said to be frozen into ice when it is converted into the solid state by cold. This conversion occurs at a temperature that is as much below the greatest heat of an Indian summer as boiling water is above. The temperature that freezes water is taken to represent the 32d degree of Fahrenheit's thermometric (*heat-measuring*) scale. Water, like all other bodies, gets denser as it gets colder (20), until it reaches the temperature of 39 degrees : below this, it expands until it freezes. Hence ice is lighter, bulk for bulk, than water, and floats upon it. " Dense " means " compressed :" it is derived from the Latin for " thick" (*densus*). " To expand " is to "open out." It is taken from the Latin word that bears this meaning (*expando*).

79. Water passes readily into the *gaseous state*, when exposed to a sufficiently high temperature.

Water is kept in the liquid state, at ordinary temperatures, by the weight of the atmosphere pressing down upon its surface (63). It bursts forcibly into the gaseous state, when it is heated sufficiently for the repulsion of its particles to have acquired force enough to overcome the resistance of atmospheric weight (24). The atmosphere rests upon a plate, one inch square, placed near the sea level, with a weight of nearly 15 pounds. When water is suddenly converted into the gaseous state, or steam, it can raise a plate of one square inch, opposed to it, with a force that it would require a weight of 15 pounds to resist. The degree of heat at which the elasticity of water is able to overcome the pressure of the atmosphere ; or, in other words, the boiling point (for water) is taken to represent the 212th degree of Fahrenheit's thermometrical scale ; the difference between the freezing and the boiling point being then divided into 180 such degrees. Water occupies 1700 times as much space again in its gaseous as it does in its liquid state. One cubic inch of water becomes about one cubic foot of steam. It

is through this conversion of cubic inches of liquid
into cubic feet of gas, that all the wonderful effects of
steam-mechanism are produced.

80. Water is, in its gaseous state, *invisible* and *transparent*, like air.

Perfect vapor, or steam, is quite invisible. The so-
called steam, that is seen to issue from the spout of a
boiling kettle, is really not steam. It is vapor partly
condensed into water, and only requiring that several
particles should be brought together, in order that they
may appear as drops.

81. Water becomes vapor at a temperature that is *considerably lower* than the boiling point of the thermometric scale.

On this account wet bodies get dry, when exposed to
the air. Their dampness evaporates or escapes as vapor.
The reason for this is, that different kinds of gases and
vapors can be diffused through each other, just as they
can through empty space (58). Each is spread under
the mutual repulsion of its own particles.

82. The atmosphere, consequently, really
consists of *a truly gaseous* (28) *and of a vaporous part* (27).

At the ordinary temperature of summer time, half a
pound out of every fifteen pounds of weight that press on
the surface of the earth, is due to watery vapor. The
other fourteen pounds and a half are due to the gaseous
portion (oxygen and nitrogen) of the atmosphere (57).

83. There is, however, generally more vapor in a *warm atmosphere* than in a cold one.

The elastic force, which causes the diffusion of the
particles of water, is more powerful in throwing them
off as vapor at a high temperature than at a low one.
If the sea were boiling, there would be 15 pounds of vapor resting upon every square inch of its surface, as well
as 15 pounds of air. When it is at 60 degrees of heat,

there is half a pound of vapor resting on each square
inch; when it is at 38 degrees, there is a quarter of a
pound.

84. Increase of temperature *increases the
load* of invisible vapor that is sustained in
the air.

The clear transparent air of a warm summer-day contains really more water in it, than the thick misty air of
a cold winter-day; but it then escapes the notice of the
senses, because it is in the true gaseous state.

85. Diminution of temperature *lessens the
load* of invisible vapor that is sustained in
the air.

Clouds are formed whenever air, that is charged with
its full load of invisible vapor, is chilled to a cooler
temperature. Visible vapor is not true steam (gaseous
water), but water in an intermediate state between its
liquid and its gaseous form. Whenever air contains only
perfect steam, it is altogether transparent and clear.

86. *Rain* is water (which has been for
some time sustained in the air, in the invisible, vaporous form,) falling to the earth.

Rain falls whenever air, that is charged with its full
load of invisible vapor, is chilled down, either by trans-
portation to regions colder than itself, or by the intermingling of colder aerial currents with it.

87. Whenever air contains its full load of
invisible vapor, bodies placed in it, that are
colder in the slightest degree, get covered by
a *deposit of perceptible moisture.*

Such moisture is, under these circumstances, denom
inated dew. The temperature, to which solid bodies
must be reduced, at any time, to cause this deposit of
perceptible moisture, is called the dew-point.

88. Cloud is vapor just condensed into the

visible form, and held sustained by the *adhesive attraction* of the aerial particles of the atmosphere.

It is well known that oxygen and water exercise some decided attractive influence over each other, for a small amount of oxygen is absorbed into water, whenever the two bodies are in contact. Hence, it is deemed highly probable that particles of water can hang suspended on particles of air, just as waterdrops do upon solid bodies; and that, in this way, depositing moisture is sustained floating in the atmosphere, as visible cloud, until it becomes too dense to be so upheld, when, in consequence of its own weight, it begins to fall as rain.

89. Constantly recurring changes of atmospheric temperature lead to frequent repetitions of *falls of rain*.

Alterations are constantly being made in the temperature of the atmosphere through the varying influences of day and night, summer and winter, and differently-blowing winds. Whenever the air gets warmer, it continues to take up more and more vapor into an invisible state, until it acquires its full charge. Then any sudden chill causes some of the vapor to pass into the visible state, and to cling as mist about the aerial particles, and, finally, when it has sufficiently accumulated, to fall by its own weight to the earth. Frequent changes from cold to heat, and from heat to cold, in this way, provide for the earth its supply of genial showers.

CHAPTER V.

THE FOOD OF PLANTS.

90. WATER *is essential* to the nutrition of growing plants.

When neither rain nor dew is deposited upon the ground during long intervals of time, vegetation languishes and dies. The chief use of rain is the preservation and support of the fertility of the soil.

91. Water nourishes plants by *carrying* to them various substances that serve as *food*.

Pure water does not sustain vegetable life. Plants placed in distilled water die very soon. The rain takes up into itself various soluble matters, that it finds floating in the air and buried in the porous soil, and carries these to the absorbing rootlets of plants.

92. Two of the most important of these nutritious substances the rain finds *floating in the air*.

Rain water always contains minute quantities of soluble matter, even when it falls fresh upon the earth. This matter it washes out of the atmosphere, as it falls through it.

93. One of these nutritious substances may be procured from rain water, *through the influence of lime*.

If a few drops of clear lime water (that is, water holding a little lime in solution) be added to rain water, and the mixture be then still further exposed to the air, it becomes slightly milky, and, after a time, a small quantity of white powdery substance settles to the bottom, which may be separated from the water by decanting or filtering the liquid off. This powdery substance is then found to be simply *chalk*.

94. Chalk consists of *lime, and another body* chemically united to it.

It has been seen that the solution of lime absorbs some other principle from the rain water and air, and that the lime then becomes chalk. The difference between the lime and the chalk is, that the former is soluble to a certain extent in rain water, and that the latter is insoluble and at once separated from it; first as a milky diffusion, and then as a powdery deposit.

95. The second principle contained in chalk may be separated from the lime, with which it is combined, by the *addition of a strong acid*.

If a little powdered chalk be placed in a bottle with some water, and a few drops of oil of vitriol, or even vinegar be added, effervescence takes place, and a clear transparent gas bubbles up through the liquid, and overflows at the mouth of the bottle. This gas it was that was combined with the lime in the chalk.

96. Chalk consists of lime and a gaseous substance, to which the chemist has given the name of *carbonic acid*.

Carbonic acid is the fixed air which escapes from soda-water and effervescing wines, so soon as the cork is drawn from bottles containing these liquids. It is driven off from its combination with lime, when chalk and a strong acid are brought together, because the strong acid then takes the lime to itself, and sets the weaker agent, the carbonic acid, free.

97. Carbonic acid is a *gaseous substance*, at ordinary temperatures; but it is easily distinguished from all other kinds of gases.

The carbonic acid, that is separated from chalk by the action of an acid, bubbles up through water, and may be collected in an inverted bottle filled with this liquid. It then displaces the water and occupies its position. Under such circumstances, it is found to be an invisible transparent substance, not to be distinguished from air by the eye.

98. Carbonic acid gas *extinguishes flame*, and *destroys animal life* immersed in it.

This property is at once exhibited if a piece of burning taper is plunged into a bottle of the gas. In this way carbonic acid is distinguished from air and from oxygen (33).

99. Carbonic acid is *very heavy* for a gas, *cannot* be made to *burn*, and has an *acid taste*.

It is so heavy that it may be poured through the air, from one bottle to another, like water; and a jet of it blown across a flame does not ignite; in these ways it is distinguished from nitrogen (54) and hydrogen (73). Water dissolves it, and acquires a pleasant acid flavor from its presence. By this means it is distinguished from another gaseous substance, presently to be named.

100. Air always *contains* a certain small proportion of *carbonic acid*, floating amidst its particles.

But this proportion is very small. There is rarely more than one part of carbonic acid to every two thousand parts of the atmosphere. The rain gets its carbonic acid from the air, and is consequently always removing this principle from the great atmospheric reservoir. The carbonic acid is, however, restored to the air, from other sources, as fast as it is removed by the rain.

101. Carbonic acid is a *compound substance*,

(37) and not a simple element, like oxygen, nitrogen, or hydrogen.

Carbonic acid may be made by combining its elements together. If a piece of red-hot charcoal be plunged into a jar of oxygen gas, atoms of the charcoal combine rapidly with atoms of the oxygen, throwing off a shower of brilliant sparks, while in the act of union (47). The oxygen gas disappears, and carbonic acid gas presents itself in its place, and may be then recognized by the properties named above (97). Carbonic acid is, indeed, solid charcoal dissolved in oxygen gas, and rendered gaseous itself by the union. When a piece of charcoal is burned, it disappears, because it is carried off piecemeal by oxygen furnished from the surrounding atmosphere.

102. Charcoal is itself an *elementary substance*, and cannot be, by any means, resolved into simpler components.

Charcoal is always a solid body, at ordinary temperatures (22). It is present in large quantities in all organized structures. It is indeed, the main solidifying agent of organization. As oxygen may be spoken of as the *energetic* element (35), nitrogen the *inert* element (55), and hydrogen the *inflammable* element of organic nature, so charcoal may be distinguished as the *solidifying* element.

103. Charcoal is called *carbon* in the language of science.

" *Carbon*" is derived from the Latin word for a coal (*carbo*); and as the acid gas, formed when charcoal is burned in the presence of oxygen, is this charcoal, or carbon united with oxygen, it has been termed *carbonic acid gas*. Wood, and all other vegetable substances are, in the main, composed of oxygen, hydrogen, nitrogen and carbon. When they are exposed to a great heat, the three gases are first driven off as watery vapor and carbonic acid gas, and the chief of the carbon is then left pure, for a time, as a black mass. If the burning be continued after their dissipation, the carbon, too, is converted into gaseous carbonic acid, and thus escapes. Coal is simply vegetable matter, that has had the greater portion of its elementary gases driven off during the lapse of ages, a certain quantity of the inflammable hydrogen only being yet left in combination with the dense carbon.

104. A black square, thus formed , may be conveniently used, *as a symbol*, to represent to the eye an ultimate atom of the solid element, carbon.

It will be remembered, as in the case of the other elements, that this black square is merely an artificial symbol, adopted for convenience of illustration (48).

105. A black square, placed between two white triangles, thus ◁█▷ , then becomes the appropriate symbol to represent a distinct *atom of the compound, carbonic acid.*

Each atom of carbonic acid contains in itself one atom of carbon and two atoms of oxygen, all united together in some unknown plan of arrangement. When charcoal is burned and turned into carbonic acid, two atoms of oxygen seize upon each atom of dense carbon, and fly away with it, so to speak, into the gaseous condition.

106. Carbonic acid can be resolved into its constituent elements only by resorting to substances possessed of the most powerful affinities for oxygen, such as the metals potassium or sodium.

Water or even ice can be readily decomposed by these metals: the oxygen unites with the potassium or the sodium, to form potassa, or soda, and the hydrogen is set free. If CARBONIC ACID is to be decomposed, the potassium, or the sodium must be made nearly red-hot, then they will rob it of its oxygen, producing potassa or soda, and the *carbon* of the gas will be deposited in its ordinary dark form.

107. Nature is constantly effecting this decomposition in the green tissues of the plants by the agency of the great chemist, the SUN, at common temperatures.

Green leaves, when placed in sunshine, are constantly at work reducing carbonic acid into carbon and oxygen. They retain the dense carbon for their own constructive uses, and throw off the gaseous oxygen, pure, into the air. If green leaves are immersed in a solution of carbonic acid, and then placed in sunshine, bubbles of the gas may be seen clinging about their surfaces as it is disengaged therefrom.

108. Carbonic acid is poured into the air wherever *carbonaceous bodies are burning*, or wherever *animals live* and breathe.

Carbonic acid gas is a constant product of the burning of carbon, in whatever form of combination it may be. It is carried up the chimney from every coal or wood fire, and it is exhaled from every breathing lung. It would accumulate in the atmosphere, in consequence of the abundance of its supply from these sources, until it reached a noxious amount, if it were not for the counteracting influence of rain and vegetable life. By this influence, it is abstracted from the air, as rapidly as it is produced; so that the medium proportion of about one part in two thousand is steadily and permanently preserved (100).

109. Dense carbon is thus made *easily transportable* from place to place, through the peculiar affinities of oxygen for its atoms.

Oxygen serves as the carrier of carbon, in the strictest sense of the term. Growing plants need carbon, as one of the most important elements of their structures; but they are fixed to the soil, and unable to go in pursuit of what they want, as animals can. Hence, the essential

substance is brought to them, by this singularly beautiful contrivance. All-present, energetic oxygen rushes
into the lungs of animals, the furnaces where fuel is consuming, and various other places, in which the processes
of decay are going on. In all of these, it seizes upon the
loosening atoms of the dense solid, and converts them,
by a touch, into a subtle, elastic gas, which is wafted
away by the passing breeze, or dissolved in the descending shower, and so finds a ready entrance into the smallest
vegetable pores. When once the compound gas is received into the interior of the plant, the oxygen is separated from the carbon, and driven off by the mysterious
influence of the solar rays, leaving the solid element to
associate itself with the other principles there present.
The very small proportion of carbonic acid, contained in
the air, is amply sufficient for all the needs of vegetable
life, because of the vast extent of the atmosphere, and
because new supplies of it are furnished as fast as it is
removed.

110. The second principle of a nutritious
nature (92) may be procured from rain water,
through the influence of sulphuric acid.
　　If a few drops of sulphuric acid be added to some gal-
lons of rain water, and the whole be then gradually evaporated by heat, it will be found that a small quantity of
a peculiar solid substance is left behind. This solid substance is the sulphuric acid combined with the principle
that was diffused through the rain water. If the solid
substance be rubbed with a little caustic lime, the sulphuric acid attaches itself to the lime, and the principle
under consideration is set free, and escapes as a pungent
vapor, readily detected by the sense of smelling. This
vapor it was, that was combined with the sulphuric acid
in the solid residue left by the rain water.

111. This pungent vapor, procured from
rain water through the influence of sulphuric
acid, is a gaseous substance, called *ammonia*
by the chemists.

The name "ammonia" was originally conferred upon this principle, because it was first noticed in the immediate neighborhood of a temple in Lybia, consecrated to Jupiter Ammon. The pungent vapor was there found, formed into the salt, known as Sal Ammoniac, by combination with a peculiar acid.

112. Ammonia is a *very volatile gas*, at ordinary temperatures, and is at once recognized by its penetrating smell.

The pungent smell of hartshorn is due to ammonia, which is constantly escaping from that liquid, in a gaseous form. "Volatile" means flying away: it is taken from the Latin word for "to fly" (*volo*).

113. Ammonia is a *very soluble gas*.

Hartshorn is merely ammonia dissolved in water. Water can dissolve its own bulk of carbonic acid gas, but it can take up 670 times its own volume of ammonia. The gas, however, constantly escapes from strong solutions, exposed to the air, in consequence of its high volatility. The pungent smell of hartshorn is due to the spontaneous escape, in this way, of the ammonia therein held dissolved. The very small proportion of ammonia ever present at one time in the atmosphere, is the result in a certain measure of its high solubility. The gas is dissolved away by the rain, and carried to the earth, as fast as it is thrown into the atmosphere. The proportion scarcely ever amounts to more than one part in one hundred thousand, and is quite inappreciable by the senses, until the process above described is employed in collecting it; still, as new quantities of ammonia are constantly in the act of being carried into the air, and of being washed thence into the soil by rain, a very large quantity of it, upon the whole, is conveyed to the ground. It has been shown that thirty pounds' weight of ammonia, at least, are carried down to a single acre of land, in the course of a year. The quantity of carbonic acid carried down by the rain and wind, is probably thirty times as much as this.

114. Ammonia is very *acrid*, and possesses *alkaline* properties.

It has been seen that carbonic acid is an acid gas (99), and unites chemically with other oxygen compounds that are not acid, to form neutral ternary compounds (45), as instanced in the composition of chalk, which is thus composed.

Chalk $\begin{cases} \text{Carbonic acid} \\ \text{Lime (itself an oxygen compound,} \\ \qquad \text{or oxide of calcium.)} \end{cases}$

Ammonia, on the other hand, is a non-acid compound, and is capable of combining with acids, to form more complex neutral compounds. It may, indeed, be made even to take the place of the lime, in the above arrangement, so that a substance thus formed is produced:

$\left. \begin{array}{l} \text{Carbonic acid} \\ \text{Ammonia} \end{array} \right\}$ Combine to make the salt called carbonate of ammonia.

The solid substance procured from rain water by sulphuric acid (110) is of a similar nature to this. It is called sulphate of ammonia, and is thus formed:

Sulphate of Ammonia $\begin{cases} \text{Sulphuric Acid} \\ \text{Ammonia.} \end{cases}$

When this is mixed with lime: 1st—The sulphuric acid is attracted to the lime, and forms with it sulphate of lime: and, 2dly—The ammonia escapes as a free gas.

115. Ammonia is a *compound gas*, like carbonic acid (101).

If a quantity of gaseous ammonia be heated to redness, in a porcelain tube, it becomes of twice its original bulk, and is then found to be two distinct gases mingled together, instead of one.

116. Ammonia consists of *hydrogen* gas and *nitrogen* gas, combined together in the proportion of three parts of the former to one of the latter.

It is found that, after ammonia has been heated to redness in a porcelain tube, one-fourth of the expanded product is pure nitrogen, and three-fourths are pure hydrogen.

117. The appropriate *symbol* for the atom of ammonia is, therefore, a black triangle surrounded by three dotted triangles, thus:

As in the case of water and carbonic acid (75–105), this is merely an artificial symbol, adopted for the convenience it affords of representing the nature of the combination.

118. Ammonia *cannot be formed* by the chemistry of man, out of its constituent elements, although it *can be resolved* into its elements by artificial means (115).

Ammonia is only to be procured through the destructive decomposition of organized structures. Whenever animal bodies are decomposed, after death, ammonia is invariably set free. It is only when atoms of hydrogen and nitrogen are in the act of disengaging themselves from union with other kinds of atoms with which they have been combined to form complex organic substances (a condition which is termed, by the chemist, the "*nascent*" state, from the Latin, *nascor*, to be born), that the inertness of the latter can be sufficiently conquered to make them contribute to the formation of this compound. It is a remarkable fact that, of these two compound gases, carbonic acid and ammonia, the former can be artificially formed, but cannot be artificially unmade again; whilst the latter cannot be artificially formed, but can be artificially resolved into its elements. Nature only can resolve carbonic acid, and nature only can make ammonia. The reduction of carbonic acid into carbon and oxygen, and the formation of ammonia out of nascent hydrogen and nitrogen, are yet among her mysteries; they cannot be imitated by the operations of man.

119. Most of the compounds formed by uniting ammonia with an acid are *fixed*, and *not volatile*.

When ammonia is carried down to the earth by the rain, it is chemically combined with acids that it finds there in the soil. The compounds are then of the nature of the fixed sulphate of ammonia, described in 114, and cease to be volatile. In this way a fixed store of the volatile gas is kept constantly on hand in the ground. Without such an arrangement, the volatile ammonia would fly back again into the air as soon as it had been deposited in the ground.

120. The compound formed by the union of carbonic acid and ammonia *is volatile*, like the pure ammonia itself.

This volatile salt of ammonia is called the *carbonate* of ammonia. It is the white substance known as smelling salts. If a small piece of it be left exposed to the air during a few hours, it is gradually dissipated, and lost to sight. In carbonate of ammonia, the chemical union is of so slight a nature that the property of the ammonia is not lost in consequence (40). The compound is pungent and volatile, like the ammoniacal gas.

121. Carbonate of ammonia contains within itself *all the four* elementary bodies hitherto named.

Its carbonic acid consists of carbon and oxygen, and its ammonia of hydrogen and nitrogen. The composition of carbonate of ammonia may be expressed in the following way :

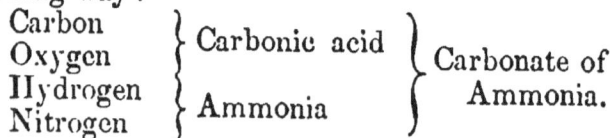

$$\left.\begin{array}{l}\text{Carbon} \\ \text{Oxygen} \\ \text{Hydrogen} \\ \text{Nitrogen}\end{array}\right\}\begin{array}{l}\left.\begin{array}{l}\text{Carbonic acid} \\ \\ \text{Ammonia}\end{array}\right\}\end{array}\left.\begin{array}{l}\text{Carbonate of} \\ \text{Ammonia.}\end{array}\right.$$

This compound, ammoniacal salt, is consequently a quaternary substance (*containing four different elements*) rather than a ternary one (149).

122. The elementary bodies, OXYGEN, NITROGEN, HYDROGEN, and CARBON, are *the most important* of the materials that are concerned in the formation of plants and animals.

If all the oxygen, nitrogen, hydrogen and carbon were taken away from the largest forest tree, there would remain nothing but a few grains of ash. If all the oxygen, nitrogen, hydrogen and carbon were removed from the body of an elephant, there would remain nothing but its bones (which are hardly vital parts), and a few additional grains of ash. Hence these four bodies have been termed "*the organic elements*" (or organogens) of nature. ("*Organogens—generators of organic structure ;*" from the Greek "*organon-gennao.*")

123. Living bodies can, therefore, get nearly all the materials for their bodies *out of carbonate of ammonia.*

Carbonate of ammonia, it will be remembered, is carbonic acid and ammonia. Water serves as a solvent and carrier of carbonic acid and ammonia, but it is also capable, to a certain extent, of giving up its own elements for the service of organization.

124. Plants form these organic elements *into organizable structure.*

They receive them as carbonic acid, water, and ammonia. Carbonic acid is introduced, in the gaseous state, by the pores of the leaves; and carbonic acid, ammonia, and water are introduced, as a mixed solution, into the pores of the roots.

125. Carbonic acid, water, and ammonia, may hence be termed the principal *food of plants.*

These three binary compounds are, indeed, the sources whence plants procure the substances they require for the fabrication of their complex organic products. They are all abundantly supplied to growing leaves and roots by the winds and rain.

CHAPTER VI. .

THE SOIL.

126. ORGANIZED structures contain small quantities of *other principles* in addition to oxygen, hydrogen, nitrogen, and carbon (122).

Wood consists principally of oxygen, hydrogen, and carbon. When a piece of wood is burned, these elements are dissipated as vapor and gas; but there then remain behind a few grains of a fixed, light-colored ash, which refuses to burn.

127. These supernumerary principles are, however, subordinate, and *not essential*, elements of organization.

Different individuals of them are found in different kinds of structure. Oxygen, hydrogen, and carbon exist in all vegetable and animal textures, and nitrogen is present in most of them; but each different kind of organic substance has a different kind of ash. The bones of animals contain lime; straw contains flint; and wood contains potash. Bone is, however, an exceptional illustration of the employment of a subordinate ingredient amidst the essential ones of organic structure; four-fifths of the substance of bone are made up of a salt of lime. Other structures do not contain more than one or two parts of ash in each fifty parts of the essential ingredients.

128. The subordinate and special ingredients of organized structures are all *fixed principles* of a mineral nature.

None of them are volatile, or (with one exception) capable of existing as gases (24) at ordinary temperatures of the earth's surface. But all of them are capable of being dissolved in water, when placed in certain states of combination with oxygen.

129. The subordinate and fixed ingredients of organized structures *are procured from the soil.*

They are formed into certain soluble compounds in the ground, ("soil" is taken from the Latin word "*solum*" the ground) and are then dissolved in the rain water, which is drunk up by the roots of growing plants. Plants get their fixed, mineral ingredients from the water, which they imbibe from the ground. Animals get them from the vegetable matters they consume as food.

130. The fixed, mineral ingredients of organized structures are *important elements* in their way, although they are of subordinate consequence to the essential principles (122) with which they are associated.

They are essential to the particular structure they assist to form, although not essential to organized structure, viewed in the abstract. Bones cannot be formed without lime; straw cannot be made without flint; and wood cannot be constructed without potash. But bones and straw, on the other hand, can be formed without potash, and straw and wood can be made without lime. These various mineral ingredients are, in fact, the additional substances that are worked up with the universal oxygen, hydrogen, nitrogen, and carbon, to render their combination the special texture required in each case. A small quantity of one is selected as suitable to convert the compound into one kind of structure, and a small quantity of another to make it into something of a different sort.

131. All these subordinate elements of organized structures *form binary compounds with oxygen* (37).

In their natural state in the soil, they are always in combination with oxygen (38). They are only obtained pure and uncombined, when they have been separated from the oxygen by the skill and ingenuity of the chemist.

132. Three of these fixed subordinate elements form remarkable *acid compounds* when combined with oxygen (45).

The three *acid-forming* subordinate elements are PHOSPHORUS, SULPHUR, and SILICON. These are all true elementary bodies; that is, in the present state of chemical science, they cannot be resolved into simpler substances. The acids which they form with oxygen, are called respectively, *phosphoric*, *sulphuric*, and *silicic acids*. The two former are made whenever phosphorus and sulphur are burned in contact with air (which contains oxygen). The third exists naturally in certain mineral substances; rock crystal is an impure form of silicic acid, and opal is the same, nearly pure.

133. Seven of these fixed subordinate elements form remarkable *non-acidulous* (or *oxide*) compounds, when they combine with oxygen (45).

The *oxide compounds*, formed by the union of these seven elements with oxygen, are—1. POTASH. 2. SODA. 3. LIME. 4. MAGNESIA. 5. ALUMINA. 6. OXIDE OF IRON. 7. OXIDE OF MANGANESE. The elements which the chemist procures from these oxides, when he has deprived them of their oxygen, are respectively—1. POTASSIUM. 2. SODIUM. 3. CALCIUM. 4. MAGNESIUM. 5.

ALUMINUM. 6. IRON. 7. MANGANESE. The last two are metals; the first five of a metal-like nature, but with the metallic characters less distinctly marked. The oxides 1 and 2 are classed together as *alkalies* (from *al kali*, the old Arabian name for potash); the oxides 3 and 4 are classed together as *alkaline earths*. (Lime and magnesia are native earths, possessing some of the properties of alkalies.) The oxide 5 is also found in a native state. The ruby, emerald, amethyst, and sapphire are nearly pure alumina.

134. Alumina forms ternary compounds by uniting with acids; but, in the absence of strong acids, it is also capable of entering into ternary combination with some of the binary compounds of the *non-acidulous, or oxide* series (45).

Alumina is very peculiar in its chemical characters. It seems to hold an intermediate position between the acid and alkaline compounds of oxygen (132, 133). The several inorganic compounds that are contained in the soil may therefore be conveniently classed, in accordance with their chemical properties, in the following way:

Acid Series.	*Oxide Series.*
Phosphoric acid	Potash and soda
Sulphuric acid	Lime and magnesia
Silicic acid	Oxides of iron and manganese.

Intermediate Constituent.
Alumina.

135. Alumina unites readily with *silicic acid* to form a ternary compound.

Such ternary compound is then chemically termed a silicate of alumina. CLAY is of this nature. It is merely a combination of silicic acid, alumina, and a certain proportion of water. When these substances are mingled together under certain favorable circumstances, clay is . formed.

136. Alumina nevertheless *leaves silicic acid* to combine with either *magnesia, lime, soda, or potash.*

When weak solutions of these compounds are filtered through broken pieces of clay, the compounds are taken away from the water, and retained by the alumina, and the water runs away pure.

137. Alumina has a *stronger affection for ammonia,* than for either magnesia, lime, soda, or potash (111).

When a solution of ammonia is filtered through clay, containing as much of those earthly and alkaline compounds as the alumina can hold, they are set free, and the ammonia takes their place with the alumina.

138. Alumina has also the power of *retaining* in its substance a certain proportion of *free water.*

Every one knows how tenaciously clay retains moisture.

139. Alumina thus *renders the soil tenacious* of the principal matters that are nutritious to growing plants.

Growing plants require, for the nourishment of their several tissues, ammonia, water, magnesia, lime, potash, and soda. All these the alumina of clay keeps mixed up in its substance (136–138). As, however, ammonia is the most volatile, but at the same time the most essential of all these substances, alumina is made more tenacious of it than of the other compounds. The oxides of iron and manganese are also commonly present in clay.

140. Plants acquire the ammonia which is essential to their vital actions (123) *from the alumina of the soil.*

In every million parts of air there is always more

than *the tenth of a single part* of volatile ammonia (113).
This the tenacious alumina of the porous soil is always
attracting to itself. Small as the proportion is, it is
nevertheless amply sufficient for all the wants of growing plants (113), for it has been shown that soils may be
kept fertile through indefinite periods of time without
any manuring, if they are but sufficiently opened out, by
culture, to enable them to draw down continued supplies
of ammonia out of the air. It is a remarkable fact
that although alumina is so tenacious of ammonia, it
nevertheless gives up a certain quantity of its prize
to pure water. Every gallon of rain water robs the
alumina of the soil of a single grain of ammonia. It is
by these means, then, that ammonia is introduced into
the interior of plants. It is first attracted from the air
by the alumina of the ground, and then washed out of
the alumina, and into the rootlets of growing plants, by
rain water that finds its way there.

**141. Lime is as tenacious of *carbonic acid*
as alumina is of ammonia.**

Whenever quick-lime is exposed freely to the air, it absorbs therefrom moisture and carbonic acid, and is converted by them into the mild compound *carbonate of
lime*. Chalk is this mild compound (94). Mortar hardens
when exposed to the air, because the caustic lime, of
which it is composed, is turned into the carbonate, by
the absorption of carbonic acid. Most soils contain
naturally more or less lime in their composition. *Marl*
is merely an admixture of clay and carbonate of lime.

**142. Carbonic acid separates from lime
whenever any of the *stronger acids* are
brought into communication with the carbonate (95).**

Carbonic acid is, chemically speaking, one of the weak-
est acids known, and therefore nearly all other acids are
able to disengage it from the calcareous compounds mingled so abundantly with the soil.

143. All productive soils contain *mould* **as well as clay and lime.**

This mould is formed of the decaying organic matters, which have fallen to the ground, and mingled with its substance, after the death of the living structures of which they have been a part. A vast diversity of textures is added to the soil in this way, but these are all principally composed of the four essential elements, oxygen, hydrogen, nitrogen, and carbon (122). Mould is, therefore, mainly compounded of these elements.

144. The mould of the soil consists chiefly of two parts. One in which *traces of organized structure are still visible;* **another in which all such traces have disappeared.**

The first part is made up of fibrous fragments that have not been yet destroyed by decomposition. The rest is a dark brown substance, that has lost all fibrous appearance, and that is in fact the fibrous part reduced into this state by being thoroughly decomposed.

145. The thoroughly decomposed portion of mould is called, by the chemist, *humic substance.*

"Humic" is taken from the Latin word "*humus*," which signifies the ground. Humic substance is the first form of simplified condition, in which the four essential elements of organized structure are found after they have been freed from the rule of life, and restored to the domain of uncontrolled chemical power. The presence of caustic lime in a soil quickens the conversion of fibrous matters into humic substance.

146. Humic substance is always possessed of *acid properties.*

On this account, it dissolves readily in a mixture of the strong alkalies and water. It is immediately separated from its combination with the alkalies, when stronger acids are added. It is quite insoluble in pure water.

147. Humic substance is composed of definite combinations of *oxygen, hydrogen* and *carbon.*

When it is first formed out of the fibrous matters, each complex atom seems to contain within itself forty atoms of carbon, fourteen atoms of hydrogen, and twelve atoms of oxygen. Its composition might, therefore, be symbolically represented in the following way.

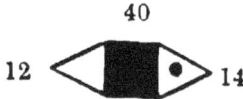

There is no nitrogen in humic substance, because the entire quantity of this element, that is separated from decaying structures, remains united with hydrogen as ammonia, and gets fixed as such in the alumina of the soil. Humic substance, on exposure to the air, continues to lose carbon and hydrogen, which are converted into carbonic acid and water, and to acquire oxygen. It is finally entirely resolved into carbonic acid and water ; this is the last stage of that process of decomposition of which the formation of humic substance is the first.

148. Productive soils, therefore, always contain *humic substance, carbonic acid,* and *ammonia,* in addition to their mineral ingredients.

Hence the chemical arrangement of the constituents of a fertile soil (134) may be extended in the following way :—

Acid series.	*Oxide series.*
Humic Substance,	Ammonia,
Carbonic Acid,	Lime,
Silicic Acid,	Magnesia,
Phosphoric Acid,	Potash and Soda,
Sulphuric Acid.	Oxides of Iron and Manganese.

Intermediate constituent capable of acting as either acid or oxide.

Alumina.

Neutral constituent capable of acting as neither.
Water.

In addition to these, there is also in the soil another
element called chlorine, which generally stands aloof
from the oxygen compounds, but which is capable of
taking the place of the oxygen in them under special
circumstances. Thus common salt, found native in the
earth in some places, is a compound of chlorine and
sodium (*chemically speaking, a chloride of sodium*). It
is soda with the oxygen replaced by chlorine. Chlorine,
in its simple state, is a gas of a greenish hue : hence it
is called by a term derived from the Greek for "green"
(*chloros*).

149. **The various constituents of the soil
are mostly in the condition of *ternary neutral compounds.***

Members of the oxide series are chemically united
with members of the acid series. When this has been
effected, the ternary body formed ceases to be like
either of the binary compounds that are united in it
(40). It is neither acid, nor alkaline, and becomes what
is termed a *neutral salt* (45). Each respective neutral
salt is called by a *generic* name that is derived from the
acid forming it, and by a *specific* name that is taken
from the oxide. Thus the salts of humic acid are called
humates, and those of the other acids, *carbonates, phosphates, sulphates* and *silicates*.* "Generic" means "of
one kind :" it is taken from the Greek word for kind
(*genos*). "Specific" means having one fixed form : it
is derived from the Latin word for form (*species*). The
leopard and the tiger belong to the same genus, or are
of one "kind ;" but they are of different species or form.

---

* The neutral salts, formed by the union of acids with alkalies, are
generally called *quaternary* compounds, because both acid and alkali
contain two elements. The term *ternary* is here used in preference,
whenever ammoniacal salts are not under consideration, because it
expresses that there are only *three kinds of elements* concerned in the
production of these neutral salts. *Oxygen* is an element of both the
acid and alkaline constituent.

150. Some of the neutral salt compounds are *soluble*, and others are *insoluble* in water (77).

Powdered chalk settles from water in which it has been diffused by shaking, because it is insoluble. But powdered carbonates of soda and ammonia remain invisibly dissolved away in the fluid, under like circumstances.

151. Most of the neutral compound constituents of the soil, being *insoluble*, are not at once washed away by the rain that percolates through the ground.

It is by the formation of these insoluble compounds that the supply of nutritious matter is husbanded, and kept stored in the ground until it is required.

152. Some of the neutral compound constituents of the soil being *soluble*, are taken up by the rain water, and are transported by it into the interior of growing plants.

The soluble salts that are thus taken up from the soil by rain water are *all the humates*, excepting that of alumina; *all the sulphates*, excepting that of lime, and the *carbonates* and *phosphates* of *potash, soda* and *ammonia*.

153. A very *small proportional part* of the soil is soluble at any one time.

If a quantity of soil be well stirred up with water, and allowed to settle, not more than two per cent. of it will be found to have remained in solution. This two per cent., however, it is that contains all the nourishment that is destined for the immediate support of vegetation. As this is removed from the soil, more of the fixed constituents are reduced into the soluble condition, to take its place. All nature's chemical operations, carried on in the ground, are directed to the double end of keeping a fixed store of nutrient material on hand, and of converting just so much of this fixed store as is required for use into a soluble state when it is needed.

154. Fertile soils must be *permeable and porous*, as well as tenacious.

The nutritious matters consumed by plants, are carried into their rootlets by water; the soil must, therefore, be permeable to water, in order that this liquid may be able to search out the nutritious matter it has to take up, for conveyance into the plant. "Permeable" means allowing any thing to pass through or penetrate : it is taken from the Latin word for to "pass through" (*permeo*).

155. Soils are *kept porous* by the presence in them of a certain proportion of *sand*.

Productive soils contain sand, as well as clay, lime, and mould. The sand, however, has no fertility in itself. Sands, that are devoid of clay and mould, are particularly barren. Sand consists of rock crystal, quartz rock, and flint (all forms of silicic acid), pounded and ground by the action of moving water into small irregular grains, which are perfectly insoluble, without adhesiveness, and of all sorts of shapes, and therefore peculiarly well fitted to make the substance with which they are mingled of a loose and permeable character. The most productive soils contain mould, clay, lime, and sand; the mould renders them rich in the organic elements (*organogens.* 122); the clay and lime make them tenacious of nutritive principles; and the sand keeps them open for the free passage of water, and for the . free admission of air.

156. *Eleven subordinate elements* are taken from the soil, to be used in the construction of organized bodies, besides the *four essential elements*, oxygen, hydrogen, nitrogen, and carbon.

Two of the essential elements, oxygen and carbon, are conveyed into plants as gaseous carbonic acid, through the pores of the leaves, directly from the air (109). But the same elements are also carried up from the soil, into the roots, as soluble carbonates and humates. All the other material elements of organized structures are taken entirely from the ground. Of the 64 different elementary bodies now known to science (45), 15 are pressed into the service of organization and used, more or less largely, in the construction of its textures. Four of these are the organogens, oxygen, hydrogen, nitrogen, and carbon; the other 11 are the bodies already specified, namely: 1. Potassium. 2. Sodium. 3. Calcium. 4. Magnesium. 5. Aluminum. 6. Iron. 7. Manganese. 8. Phosphorus. 9. Sulphur. 10. Silicon. 11. Chlorine.

CHAPTER VII.

CELL-LIFE.

157. WHEN organized structure is finally prepared for the service of life, it is moulded into the form of a *vesicle or cell*.

The term "cell" is derived from the Latin word "*cella*," which signifies "a chamber;" a cellar or place prepared for the reception of some precious store. "Vesicle" means a small bladder: the word is derived from the Latin "*vesicula*," a little bladder.

158. A cell consists of a *film*, (or delicate skin,) arranged so that it incloses a hollow space.

The term "cell" is in common use, as the name of this rudimentary form of living structure. But the word vesicle which is sometimes employed instead, is really the more appropriate term of the two, on account of expressing at once that the thing alluded to is simply a little bladder.

159. The cell-film is composed of a complex substance, made principally of the *four essential elements* — oxygen, hydrogen, nitrogen, and carbon (12£), combined together.

Chemists have not been able fully to determine the precise nature of this delicate substance, but there is great probability that it is built up of complex particles which contain each within itself forty-eight atoms of carbon, thirty-one atoms of hydrogen, twelve atoms of oxygen, five atoms of nitrogen, besides minute traces of sulphur, phosphorus, and some of the other subordinate elements. Complex particles of this kind, employed for the construction of organized bodies, are called *organic molecules* ("Molecule" means "a little mass"). Each organic molecule of cell-film may hence be *symbolically* represented in the following way.

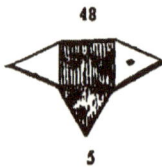

To form one of these organic molecules, 48 compound atoms of *carbonic acid* (105) are first collected, then all but 12 of their 96 atoms of oxygen are driven away, and 31 atoms of hydrogen, 5 of nitrogen, with the minute traces of sulphur, &c., alluded to above, are put into their place, and the whole are chemically united into one mass. In the construction of cell-film, several of these organic molecules are placed side by side in the direction of a thin layer, and are so cemented together. Some of the chemists have agreed to designate the organic substance, that is *first formed by living bodies*, out of the organic elements, as the basis of their cell-films, by a specific name. They call it *proteine* (from the Greek *proteuo, to take the first place.*)

OBS. *Organic* bodies are not of necessity *organized.* They are simply substances perfectly prepared to be employed in the building up of organized structure. Thus the substance used in the formation of cell-film is *organic,* but the cell-film itself is *organized.*

160. Cell-films are destitute of *perceptible pores.*

The highest powers of the microscope fail to detect any holes in them. They merely look like thin, transparent, homogeneous films ("*homogeneous*," " *of the same kind throughout*," *from the Greek, omogenes—omos,* "*the same ;*" *oginomai,* "*to be born*").

161. Cell-films are, nevertheless, *permeable by liquids.*

The vesicular cells contain liquid in their interiors; this must, therefore, have got through the cell-films in some way or other. There may, possibly, be pores in them too small to be visible to even the most powerful microscopes.

162. Organic membranes imbibe liquids, in virtue of a *peculiar property* with which they are endowed, when placed in special circumstances.

If a quantity of syrup be tied up within a bladder, and this be hung up on a nail, it will be found that none of the syrup drips through. But, if the bladder be taken from its nail, and be thrown into a pail of pure water, some of the syrup immediately begins to pass through the bladder, to mingle with the water in contact with its outer side; and some of this water, in turn, finds its way in to mix with the contained syrup. The water is sucked into the substance of the bladder (as it would be into blotting paper), until the water, the syrup, and the membrane are all in close connection, and able to influence each other.

163. The process by which organic films

imbibe liquids, through their substance, is called *osmose*.

The word "osmose" is derived from the Greek *otheo*, to push forward (*osmos*, a pushing forward).

164. The special circumstances which must be present, in order that organic membranes may imbibe by osmose, are the existence of *different kinds of liquids* on opposite sides of the membranes.

In the experiment with the bladder, its substance separates syrup from clear water. If syrup of the same thickness, or pure water alone, had been on both sides of the bladder, no passage of liquid through its substance would have occurred.

165. Osmose is an operation capable of exerting *considerable power*.

The process of osmose may be conveniently observed by tying a piece of bladder closely over one end of a wide, glass tube, as represented at T in the sketch. (*Fig. 2*.) If, then, syrup be poured into the tube, until it is about one-fourth full, as shown at S, and the whole be set into a vessel, V, filled to about the same height with pure water, and be left there for some time, it will be seen that the syrup gradually rises up into the tube, against the force of gravity, at the expense of the water in the outer vessel. It has been ascertained that, under some circumstances, a single square inch of organic membrane may be made, in this way, to lift and sustain no less than seventy pounds' weight of liquid.

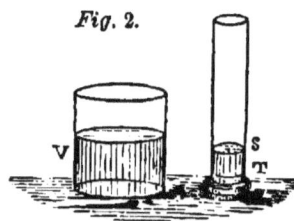

Fig. 2.

166. The operation of osmose is attended *by the destruction* of a part of the substance

of the organic film, through whose instrumentality it is sustained.

Professor Graham has recently shown that osmose is really a chemical operation. It never takes place unless organic molecules of the film are destroyed, and resolved into simpler combinations, through the influence of the molecules of the contiguous liquid.

167. Osmose is attended by the passage of a portion of the denser liquid *towards the thinner one.*

In both the experiments alluded to above, (162–165) portions of syrup pass outwards into the pure water, as well as portions of the water passing inwards into the syrup. This outward movement of the denser liquid through the bladder is called *exosmose,* or an " *outward pushing,*" (from the Greek *ex* and *osmosis*). It is, however, in no way dependent on chemical action; it is merely a mechanical process—the diffusion of unlike liquids, each amongst the other—tending towards equalization of composition. The thin fluid always moves more rapidly inwards, under the influence of osmose, than the denser one does outwards, under the influence of exosmose. Hence, in the experiments, the bladder of syrup gradually enlarges in bulk, and the column of syrup rises higher within the glass tube.

168. When osmose takes place through living structure, portions of its substance are *built up by nutrition* as fast as others are removed by chemical action.

Living structures are constantly undergoing change of internal composition. They are nourished by food, at the same time that they are wasted by the actions of life (5).

169. Liquids are altered in their charac-

ters when they are *passed through living structure*, by the operation of osmose.

The matters contained within the cavities of living cells (and which are introduced there by osmose), differ altogether in nature from the matters by which those cells are surrounded.

170. Living cells contain *viscid liquids* with granular matters floating in them.

These cell-contents are readily discerned, when the microscope is employed in viewing them, in consequence of the transparency of the cell-film by which they are inclosed. In the accompanying figure (*Fig.* 3), the microscopic appearance of a living cell, with its granular contents is shown. "Viscid" means "tenacious"—or gluey. The word is taken from the Latin term for the mistletoe (*viscum,*) which has tenacious juice. "Granular" signifies composed of minute particles or "grains." It is derived from the Latin for a grain of corn (*granum*).

Fig. 3.

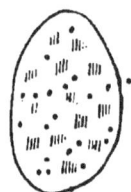

171. A portion of the substance contained within living cells, is organic matter *prepared for the support of their structure.*

The *organic elements* are combined into *organic molecules* on being imbibed through the films of living cells. The crude elements of organization, when brought together within the direct influence of living structure, are fitted, in their turn, to become living structure. Cells thus prepare and perfect their own sustenance. The precise way, in which this object is effected, is yet, however, among the hidden secrets of the operations of life. The organic substance, prepared for the support of the cell, is contained in the viscid *cell liquid.* The granular matters that float in this liquid are destined for another purpose.

172. Living cells, besides preparing the

nutrient matter of their own structure, also *form other cells* like themselves.

Each cell produces its own living offspring, which mature and become independent cells in their turn, and then also produce other bodies like to themselves.

173. Young cells are commonly produced in the interior of old ones by a *gathering together* of several of *the granular bodies* that float in the cell liquid.

This operation may be watched while in progress, if the microscope be employed. Several of the small, floating grains are brought together into one mass, and a delicate film is then raised up, blister-like, from its surface, and this enlarges progressively until it becomes a complete cell. The figure (*Fig.* 4) shows the microscopic appearance of a cell, in which a young vesicle is in process of formation in the interior.

Fig. 4.

174. Young cells *grow* by feeding upon the organic substance which they deposit in their own interiors.

Young cells may be seen to grow larger and larger, until they mature into the dimensions of the parent vessels.

175. Old cells, after they have reached maturity, and produced their offspring, *die*.

Very commonly, a brood of young cells is formed at once in the interior of one old one. The parent is then finally burst and destroyed, in consequence of its offspring growing too large for its internal capacity. The young cells escape from their parent, in this way, to begin independent life on their own account.

176 Some cells multiply by forming partitions across their cavities, and by then *splitting asunder* through these partitions.

In the figure, (*Fig.* 5,) a microscopic view is given of a cell forming its partition, preparatory to splitting across to make two distinct individuals. When new cells originate in this way, each half of the primary vesicle grows, subsequently to the division, until it has attained the dimensions of the parent, and then, in its turn, begins to form a partition in preparation for its own division.

Fig 5.

177. Some cells multiply by forming *buds* on *their outer surfaces*, which separate spontaneously after a time, and develope into mature cells.

In the figure, (*Fig.* 6,) a cell is shown, forming its young external bud. In this mode of cell production, the parent vesicle is not destroyed in the act of producing its offspring, as is the case in the other two modes described (173–176). The method of cell formation which, proceeds by the division of the parent, is called *fissiparous*, that is, production by division (from the Latin *findo*, to cleave, and *pario*, to bring forth). The method which proceeds by the formation of buds is termed *gemmiparous*, that is, production by buds (from the Latin "*gemmo*," to bud, and "*pario*," to bring forth).

Fig. 6.

178. Living cells commonly *alter their general form* and appearance as they grow towards maturity.

They do this by depositing a dense material upon their cell films, either as an inner or an outer coat. These deposits are generally figured and marked in various ways, so that the cell assumes a variegated appearance after it is thickened by their presence. In the figure, (*Fig.* 7,) *a* represents the microscopic appearance of a mature cell, thickened by a deposit of woody mat-

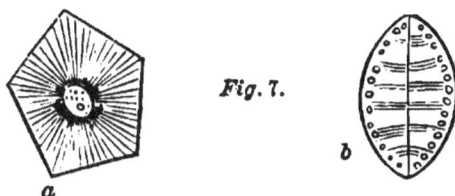

Fig. 7.

ter on the inside of·its film; *b* is a mature cell, covered by an external flinty shell.

179. A living cell is a *distinct and perfect organ* in itself.

It is an *instrument* (2) fashioned for the performance of certain definite offices, and able to accomplish those offices of its own accord, without the reception of any extraneous aid.

180. The organic offices which living cells perform are *five-fold.*

1. They make organic substances out of the crude elements of organization (171). 2. They grow by attaching portions of that organic substance to their own structure (174). 3. They vitalize the substance they attach to their structure, and enable it to perform like functions with themselves. 4. They alter their own form and appearance as they grow mature (178). And 5. They produce other living individuals like, in all respects, to themselves (172).

The functions of cell life are, then, formation of organic matter :—growth; vitalization of structure ; transmutation of form ; and reproduction.

181. A living cell is also a *distinct and complete individual* in itself.

A living cell is an *individual creature* as well as a *perfect organ.* It springs from a germ, grows, matures itself, and then reproduces its kind, and dies, just as the most perfect plant or animal does.

182. There are myriads of living creatures · that are nothing else but *isolated simple cells.*

Such creatures are called unicellular, or *one-celled* organisms. They are found in very numerous positions on the earth. Water, both salt and fresh, abounds in their forms. Every surface that has been long exposed to the conjoined influences of moisture and air, is sure to contain them. One of their favorite places of resort is the surface and bottom of still pools of water, whence they may be skimmed for examination. Entire beds of what were once deemed mineral bodies are now known to be thick layers of the indestructible remains of these curious objects. The City of Richmond, in Virginia, is built upon a foundation of them, eighteen feet deep. Many hundred species are classed together by naturalists under the family name of the *Diatoms*. The appellation is derived from two Greek words (*dia temno*) which signify " cut through," and is conferred upon them because they multiply by the method of subdivision (176). They are also termed *Brittleworts*, on account of their being generally protected externally by thin, brittle shields. The figure at *b*, in paragraph 178, is the sketch of a Diatom. These unicellular organisms are of exceeding interest to physiologists, because they are really the rudimentary forms of living creation. In them they see life in its simplest and least complex condition. While investigating the operations that the humble Diatoms perform, men of science are really ascertaining what the fundamental actions are that the complicated organizations of their own wonderful bodies are carrying on.

183. Living cells are mostly of very *minute dimensions*.

They are generally of microscopic size; and many of them require the highest powers of the instrument to be employed before they can be discerned. Thousands, and sometimes even millions of their bodies, may be contained in an area that is not larger than the surface of a shilling. The one drawn in paragraph 178 is enlarged between 300 and 400 thousands of times; that is, more than three hundred thousands of the real creatures might be packed away on the surface of paper that is covered by the drawing. A hundred of some of the Diatoms can lie side by side within the breadth of a hair. There are species of this tribe that must weigh as little as the two hundred millionth part of a grain. It has been calculated that it is possible for some million of millions of them to be contained in a cubical inch.

CHAPTER VIII.

VEGETABLE ORGANIZATION.

184. Plants are *composed of cells* (158) piled together in countless numbers, and so united into a continuous mass.

A plant is built up of cells, as a wall is built up of bricks. The individual cells of plants are generally so small that they cannot be seen by the naked eye, but there is no part of vegetable structure in which cell-formation may not be detected by the microscope.

185. Plants grow by the *multiplication of the cells* of which they are composed.

The cells, of which plants are built, multiply by the processes of subdivision and budding already described (172, 176, 177). The only difference between the results of cell-multiplication, where unicellular organisms, and where continuous masses are concerned, is that, in the one case, the young broods are cast loose as soon as they are produced, while, in the other case, they are all still held together, and connected with each other.

186. Some plants are entirely composed

of *living cells,* which continue to perform vital functions so long as the plants exist.

This is the case with all the simpler species of plants, such as the mushrooms, mosses, sea-weeds, and their fresh-water allies. The figure at *a* (*Fig.* 8) gives a mi-

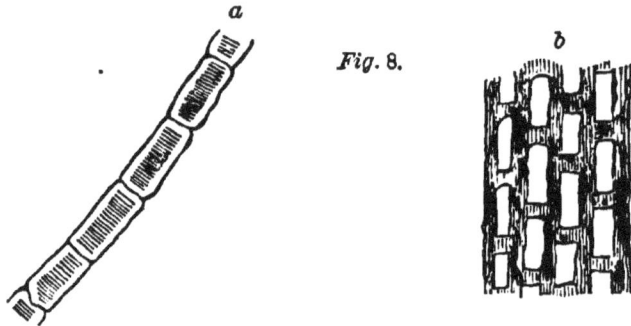

Fig. 8.

croscopic representation of the structure of a water plant, known as the Conferva. It is nothing more than a row of living cells placed end to end, and united together. This plant grows by the subdivision of each separate cell. After each division, the newly-formed vesicles increase in size until they attain the full bulk of the parent from which they have been formed, and, in this way, the row of cells is of necessity lengthened. The figure at *b* illustrates the structure of a sea-weed known as the Dictyosiphon (*net-tube*). In this several cell-rows are placed side by side, and connected together into an expanded layer. The transition from the unicellular organisms to these simplest kinds of plants, that are composed merely of rows and layers of connected, transparent cells, is a very gradual and interesting one. In some species of Diatoms the individual cells cling slightly together by their corners or edges, as if loth to part, instead of being at once cast quite asunder when completely formed. The figure at *c* (*Fig.* 9) shows the appearance of one of these connected Diatoms, known as the Bacillaria, or zig-zag Brittlewort. The yeast that rises from the fermentation of beer affords another beautiful illustration of the same kind. When examined with a powerful microscope, this substance is found to be almost entirely composed of a countless myriad of living cells. So long as the yeast is in a fermenting state, these cells multiply very rapidly by the process

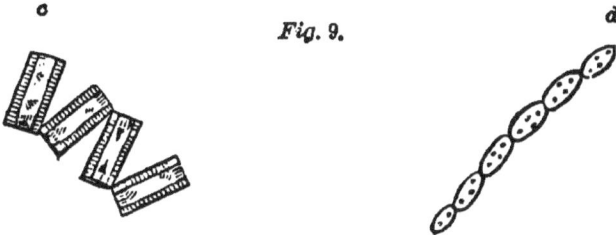

c

Fig. 9.

d

of external building (177), and then stick together in lengthened rows, something like the branch of the conferva. But the instant the fermentation ceases, the rows fall to pieces, and the cells separate, and become true unicellular organisms. The figure at *d* is the sketch of a microscopic view of the yeast cells connected in rows during fermentation.

187. Some plants are *partly composed of* living cells that are performing important vital functions, and *partly of dead cells* that have altogether ceased to fulfil any of the offices of life.

All ordinary trees and shrubs, that possess a woody texture, are of this kind. There are living cells in the leaves, but the wood of the branches and stems is entirely composed of dead ones. The old wood of trees is really no longer alive, or capable of growth, or of any of the other operations of vitality. The only vital structure in such is found in the green leaves, and in a layer of young forming tissue, that lies outside the old wood, and immediately beneath the bark.

188. The dead cells, that constitute the wood of trees and shrubs, *endure for long periods of time,* because they are filled with

D

a dense hard matter that is indisposed to decay.

Old and hard wood is entirely formed of cells that
have become altered in character subsequently to their

Fig. 10.   first formation, by the deposit of successive
layers of dense substance in their interiors
(178). This substance is of a firm, hard
nature, and does not decay, unless very
freely exposed to moisture and air. Wood
cells are long and tapering at each end, and
they are wedged in very tightly and firmly
amongst each other; hence they form a
very strong and resisting structure, even
although themselves individually dead. In
the figure (*Fig.* 10), the microscopic appearance of a
mass of wood cells is shown.

189. Cells that are designed for the continued performance of the functions of cell
life (180), remain *thin and free from any
dense deposit* upon their films.

Living cells must be to a certain extent pervious to
liquids, in order that they may keep up the operations
of osmose (163), and the transformation of crude matter into organic material (171); hence they are kept
free from the deposit of any dense and impervious
layers upon their films. The living Diatoms are covered externally by flinty, impenetrable shields, but
pores are left in certain places through these, in order
that liquids may there be readily brought into communication with the cell films.

190. Leaves are principally composed of
thin-walled and living cells, actively engaged
in the business of cell life.

For leaves are the really living portions of vegetable
structures. The stems and branches of perennials are
principally old parts that were alive years ago, but that

are now dead, although preserved from decay by the dense deposits they contain.

191. When cells are coated by layers of dense deposit, they are designed for the performance of *merely mechanical*, in the place of vital, offices.

This is the case with wood cells. They fulfil none of the ordinary duties of cell life (180); but the structure which they form serves to uphold the still living organs, the leaves, and to spread them out widely into the light and the air.

192. Leaves contain some *coated and dead cells*, as well as the thin living ones of which they are mainly composed.

These are arranged as a stiff frame-work, upon which the thin, living cells can be laid. A leaf consists of a skeleton of branching fibres that issue from the foot-stalks, and that are mainly made up of bundles of hard wood tubes; and of an expanded, green blade, that extends from fibre to fibre, and fills up the interspaces. This blade is composed of layers of thin-walled, living cells.

193. Wood is formed through the *agency of leaves.*

The nutritious matters that serve the plant as food, are imbibed by the rootlets from the soil, and then rise as sap up to the leaves. There the sap gets changed and condensed into various complex matters, under the influence of the operations of cell life, and among other things, into woody substance. This woody substance, as yet in a somewhat liquid state, is then sent back, along channels left for the purpose in the fibres and foot stalks of the leaves, to the stem, and is there stored away as a deposit in the older cells, until they are quite consolidated and choked up. The quantity of wood, that is formed by any tree during one season, is in pro-

portion to the amount of foliage it bears. When all the
leaves are picked off from young shoots, as fast as they
appear, no wood is formed by them. When trees have
more foliage on one side than on the opposite, they always have most wood, too, on that side of their stems.

194. When leaves have finished the productive work of a season, they *drop from the
tree*, and *decay*.

The wood cells of the foot-stalks of leaves continue to
imbibe more and more of the dense deposit that is sent
back through the channels they lie near, until at last
they get completely filled and choked up. When this
has happened, the base of the foot-stalk can no longer
enlarge, as the surface of the expanding stem, to which
it is attached, does. Consequently the stem tears itself away from the leaf-stalk, and the dead leaf falls.
Stems increase in size, because there is always a layer
of substance outside the old wood, and beneath the bark,
which is composed of young, living, and multiplying
cells, that are ready to be converted into new wood by
the reception of descending wood substance. The old
leaves decay, when they have fallen to the ground, because they are principally composed of thin cells that
have no dense deposit in them to protect them from
quick decomposition.

195. The nutritious sap ascends to the
leaves, *through ducts* or vessels, provided for
its conveyance from the rootlets.

These ducts are made out of cells. Drum-shaped
vesicles are piled up on each other in rows, and then
the contiguous surfaces are removed, so that the cavities
of the cells open into each other, and form continuous
tubes. In the figure (*Fig.* 11), the microscopic appearance of a portion of one of these sap-conveying ducts
is sketched. Ducts of this kind are nearly always found
amidst the bundles of new wood cells. Their sides are
thickened and strengthened by the same kind of deposit

Fig. 11. that is found in wood cells; but their bores are so large that they do not readily get choked up by it. When a young shoot of a perennial is cut across, the orifices of these ducts may be discerned by the naked eye. They are very perceptible, indeed, in the section of the common cane. The ascending sap also makes its way through the young wood cells, before they are thickened and lined by their dense deposit. Ducts are placed, too, along the woody fibres of leaves to serve for the conveyance of sap to the thin leaf cells, and of perfected wood-matter back.

196. New leaves are produced each successive growing season from *leaf-buds.*

Small projections may be seen standing out just above the scars whence the old leaves fall in the autumn. These are called leaf-buds, and are formed of living cell structure. They are, indeed, merely little extensions of the layer of young cell matter, that lies immediately beneath the bark, and that is destined for the formation of the new wood. They remain inactive and dormant through the winter, but so soon as the warm sun of the spring excites the forces of vegetable life, their individual cells begin to multiply, and so they get pushed forth into shoots. The figure (*Fig.* 12) represents the appearance of a young leaf-bud, projecting just above the scar from which an old leaf is fallen.

Fig. 12.

197. Every leaf-bud on a tree is a distinct and *perfect individual,* capable of maintaining an independent existence of its own.

Hence, when a slip containing a leaf-bud is cut off from a tree, and planted in the ground, it still goes on pushing out its shoot, and maturing into a leaf-covered branch. The only difference between a leaf-bud remaining attached to a tree, and one that is separated from it and planted in the ground as a cutting, is, that the one sends its roots down into the stem to mingle there with others belonging to the community; while the other sends its roots into the soil, and begins at once an altogether independent life.

198. A tree is a *collection of individuals* associated together, and structurally connected, rather than *an individual* distinct in itself.

Hence it is, that a tree can be lopped into pieces, which are then capable of taking root, and growing into mature trees on their own account. Highly organized animals cannot be divided into subordinate individuals in this way, because they are already individuals, and not connected clusters. In the plant, the *individuality* is in the several organs of which the whole is composed; in the animal, it is in the whole which is formed out of the several organs. The fact is simply that all the leaf-buds on a tree are formed exactly alike, and, therefore, any one may be removed from the rest, without being functionally missed in the system; but all the several parts of an animal are unlike each other, and are adapted each to perform a separate office in the economy from the rest, so that one cannot be removed without derangement being produced in the actions of the whole.

199. Each leaf-bud *forms its own roots*.

If two slips of willow be taken, and all the leaf-buds be picked away from the one, while they are allowed to remain upon the other, and the two slips be placed together in a glass of water, it will be observed that rootlets are, after a little time, pushed out from the bottom of the slip containing the leaf-buds, while none are formed on the other. In the little water plant, known as the Frog's Bit, rootlets may be seen descending into the water from the parts of the stem where the leaf-buds are. In the duck-weeds, they may be traced as so many little threads dropping from the leaves themselves.

200. In woody plants, roots are the downward *prolongations of the wood fibres* of the stem.

In the Banyan tree, which has several stems united together above by transverse branches, roots descend from the transverse branches and fix themselves in the ground, thus becoming stems in their turn. In the screw-pine, root-fibres push their way out from the lower part of the stem, and then pass on obliquely to the ground.

201. Roots are formed of *ducts enclosed within bundles* of *woody fibres.*

In this way, strong tubes are fashioned, which are admirably adapted for the conveyance of nutritious liquids up to the leaves, and which are yet able to protect their contents from accident.

202. Roots only admit liquids *at the extremities* of their fibres.

The several fibres of the root separate from each other in the soil, and branch out as fine threads in all directions. Each one is, however, strengthened by a sheath of wood cells. which excludes liquids every where, excepting at its extremity ; there an opening is left in the woody sheath, and a small mass of thin-sided cells is inserted, which is able to suck fluids in, that are placed in contact with its outer surface, under the influence of osmose (163). The little masses of thin-sided, absorbing cells, placed at the root-points, are appropriately called *spòngioles,* or "little sponges."

203. The absorbing spongioles of the roots have the *power of selecting* such matters from the soil, as are suitable for the nutrition of

Thus they take in at once the soluble humates and carbonates, and other salts of ammonia, lime, potash, soda, iron, &c. (148), which comprise the various materials that are employed in the fabrication of organic structure, but they reject most other compounds that come within their reach as either useless or noxious to the living plant.

204. Roots grow by the continued addition of new materials *to their points.*

The cells of the spongioles divide and multiply, and as the new progeny remain attached to the old mass, the extremity of the rootlet is advanced upon each addition. The older cells of the spongioles left behind, then get altered and filled up to constitute the dense substance of the root-fibre. One consequence of the rootlets being formed in this way is, that they are never pushed on against opposing obstacles in the soil. They creep round hard substances that chance to lie in their way, by adding new matter in front towards the direction which is open to their advance. Leaves, which are developed in the air, where there are no obstacles, are pushed on by the addition of new cell matter behind. Hence, when the points of young leaves are nipped by the frost, they are never renewed.

205. The most highly developed plants are organized by the *operation of the same powers* that are influential in calling the simple, living cell into being.

There is no active power at work in the majestic oak tree that is not concerned also in supporting the life of the simple, one-celled organism, that floats in the surface film of a drop of water, and that cannot be seen without the aid of a powerful microscope. The oak tree, indeed, is nothing more than an enormous pile of such microscopic cells and their dead remains, held together and preserved by the products of their own activity. The history of the largest plant is simply that a number of living cells multiply and produce successive generations of offspring like to themselves, and that, of these offspring, some are preserved by the alteration of their forms, and by the incrustation of their delicate films with dry, hard coats (191), and are then held together as connected masses; some are transmuted into ducts and vessels for the conveyance of nutritious matter from the soil, losing their lives, too, in the transmutation (195); while others are reserved in the living state (189), and are caused to carry on the various functions of cell-life (180), being, however, consequently obnoxious to the penalties of life, that is, subject to the unalterable laws of birth, growth, decay and death. Forest trees are monumental records of past cell life, reared out of cell remains, but covered also with a surface drapery of still existing cell life, that goes on year by year adding new incrustations to the pile within. Such is vegetable organization.

206. The cells, living and dead, of which the more highly developed plants are built, are of *extremely minute dimensions.*

The largest of them can scarcely ever be seen by the naked eye. The smallest require to be magnified by the microscope many thousands of times, before they can be discerned. About fifteen hundred of the largest might be laid out, side by side, upon the space enclosed within the four straight lines of *Figure* 13. About fifty thousand of those of average size might be packed away on the same surface; and of the cell alluded to as connected with the fermentation of yeast (186), as many as four millions.

Fig. 13.

D*

CHAPTER IX.

THE CONSTRUCTIVE OPERATIONS OF PLANTS.

207. ALL the great operations of vegetable life are carried on *in the leaves* of plants.

Leaves are principally composed of living cells (190), hence the various functions of cell life are carried on within their structures. It is in them that the work of organization is mainly performed.

208. Sap is abundantly furnished to the leaves of plants, in order that they may be *supplied with the materials* of organizing work.

Sap consists of the various nutritious principles, that are adapted to the support of vegetable structure, mingled together and dissolved in water. The spongioles of the rootlets imbibe these matters from the soil (203), and they are then conveyed up to the leaves. through the ducts and young wood of the stem and leaf fibres. When a wound is made into the young wood of trees in early spring, the rising sap is seen to flow out from the under portion of the wound.

209. Sap is conveyed from the soil up to the leaves of plants, by the *influence of osmose* (163).

Every little cell film in the fabric of the vegetable so placed that it separates two different kinds of liqu from each other, for the sap is changed in nature ?

being passed through the film (169); hence it *pushes forwards* the liquid that is in contact with one of its surfaces (163). For reasons that will presently appear, matters are so arranged in plants that this "pushing" is propagated onwards from the very spongioles of the rootlets up to the leaves. The spongioles imbibe liquids from the soil by osmose; these liquids are then conveyed from cell to duct, and from duct to cell by osmose, being changed by cell influence into sap as they go; at last the leaves imbibe the sap through their foot-stalks from the stem by osmose.

210. Plants are clothed externally by a nearly *impervious skin.*

This skin is formed of a layer of flattened cells placed side by side, and is varnished over externally by a thin covering of dried mucilage. These cells are filled with liquid inside, but they have air externally on one side; consequently as their outer films have not "unlike *liquids*" on each side of them (164), no further motion is effected by osmose when the sap has reached the outer skin-layers of the leaves. The external covering of dried mucilage effectually prevents all evaporation through the skin.

211. The external skin of leaves is pierced by a *great number of pores.*

These pores are provided to allow of the escape of vapor from the interior of the leaves. If it were not for them, nothing could get away from leaves after it had once entered their structures, unless by passing back through their foot-stalks to the stem. If a cold glass be inverted over some growing plants, or over grass on a clear day, it will be observed that its inner surface very soon gets bedewed with vapor; this has all been thrown off from the sap of the plants through the exhaling pores of their leaves. If the roots of some vigorously growing plants be placed in a glass of water, and another glass of the same size, and containing the same quantity, be placed by the side, it will be found that the water containing the plants will be wasted much more rapidly by evaporation than the other will ; this water, indeed, is pumped up into the leaves of the plants, and is then exhaled copiously from their pores. It has been ascertained that a plant of the common sun-flower exhales from twenty to thirty ounces of water every day from its leaves ; and an acre of grass land as much as 6400 quarts in twenty-four hours.

212. The exhaling pores of leaves are so constructed that they can be *opened and closed* according to the need of the time.

The exhaling pores are placed between two crescent-shaped cells, which meet by their points, and thus leave an opening bounded by their concave borders. The figure (*Fig.* 14) represents, in a highly magnified view, the flat cells of the skin of a plant, and the crescent-shaped vesicles inclosing an exhaling pore between them. Whenever the air and soil are very dry, and plants cannot afford to lose much moisture, the concave edges of the crescentic vesicles are straightened, and pressed together, and the openings of the exhaling pores in this way closed. When, on the other hand, the air and soil are laden with moisture, the inside borders of the crescentic vesicles are curved away from each other, and the exhaling pores opened. The exhaling pores are therefore really *mouths*, affording a regulated communication with the interior spaces of the leaves. They are called *stomata*, a word derived from "*stoma*," the Greek term for mouth.

Fig. 14.

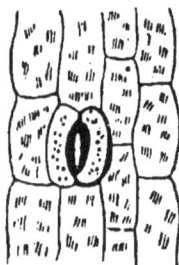

213. The exhaling pores are chiefly placed on the *under sides* of leaves.

They are so placed, because they are then removed from the direct influence of the scorching sun. They open into loose spaces left among the thin-walled leaf cells. The figure (*Fig.* 15) represents the arrangement of the cells of a leaf; the sketch is made from a thin slice cut out through the thickness. *ss* are the flattened cells of the upper skin, covered by their layer of dried mucilage : *cc* are the cells that form the general thickness of the leaf; and *pp* are the exhaling pores opening between the crescentic vesicles (seen in section), and through the underskin, into loose spaces lying amongst the thin-walled cells. Some plants have as many as 160 thousand of these little pores on every square inch of the underskin of their leaves.

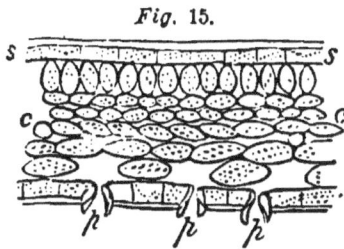

Fig. 15.

214. No exhalation takes place through the pores of leaves, unless plants are exposed to the influence of *solar light*.

When the leaves of plants are exhaling copiously, the exhalation is stopped instantaneously upon removal into darkness. Heat promotes exhalation, but it is unable to effect it without the aid of light. Light causes the crescentic vesicles to curve in their sides, and so to open the intervening pores. When the pores are thus opened, more moisture of course passes through them, under the influence of simple evaporation, in any given time, if the air be hot and dry, than if it be cold and moist. Light opens the pores of plants, and heat causes increased exhalation through the opened pores.

215. The vapor that is exhaled from the pores of leaves is water *very nearly pure*.

Fifty ounces of the water thus exhaled do not contain so much as a single grain of any other substance.

216. The exhalation effected in the leaves merely *thickens the sap* that is exposed to

the influence of light in the cells of those organs.

When water is driven off from the sap contained in the cells of leaves, since nothing else is taken away with it, the remaining liquid is left the thicker for the loss, just as syrup is thickened when a portion of its water is driven off from it by evaporation over a fire. This is one great change, then, that is brought about in sap after it has been poured out into the leaf cells; it is thickened by the evaporation of its thinner parts through the exhaling pores, under the influence of exposure to the light and heat of the sun.

217. Leaves *exhale true gaseous matter*, as well as watery vapor, so long as they are exposed to sunlight.

If a few green leaves are placed in a plate of water acidulated with carbonic acid, and this be then left in strong sunshine, small bubbles of gas will soon be seen clinging about the under surfaces where the exhaling pores are situated (107). These bubbles are clearly not formed of watery vapor, because, if they were, they would be condensed in the water, and altogether disappear.

218. The gas which leaves exhale in sunshine is *pure oxygen* (34).

If a quantity of these bubbles be collected under water, in a small glass tube, and be then subjected to examination, it will be found that the gas possesses all the characters of simple oxygen (43–47).

219. Oxygen gas is not exhaled by leaves *in darkness*.

The removal of a growing plant into darkness stops the exhalation of oxygen gas from its tissues, as well as that of water (214).

220. The oxygen gas exhaled by leaves

is procured from the *carbonic acid* that is introduced as food.

If the green leaves (217) are placed in pure water, that is entirely devoid of the presence of carbonic acid, no bubbles of oxygen gas are formed when they are exposed to sunshine. If two glass jars be inverted in a water bath, as represented in the drawing (*Fig.* 16), one of them, W, being also filled with water, and the other, C, containing only carbonic acid gas, and if two or three plants of water-mint be introduced into the jar W, and the apparatus be then placed in strong sunshine, it will be found that carbonic acid is gradually taken from the jar C, through the water in the bath B, to the plants growing in W, and that oxygen is then given out by the plants into the jar, displacing the water at its top, and gradually accumulating there. As this occurs, the water also rises to the same amount into the jar C, to take the place of the carbonic acid removed thence.

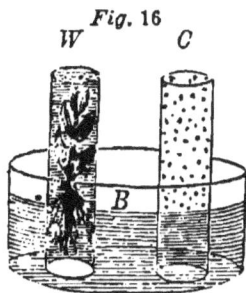

221. The carbon of the carbonic acid introduced into leaves *is used in the formation of organic material.*

This is the second great change that is effected in sap, when it is exposed to sunlight in transparent leaf cells. Its carbon is mingled organically with other elements, and oxygen is set free into the loose spaces, to be thence exhaled with the watery vapor, through the stomata. One of the most important offices, performed by living vegetable structure, is the fixation of the great agent of solidification, carbon (102), into complex, organic substance. It is thus that green leaves effect what the science of man cannot accomplish, the separation of the constituent elements of carbonic acid (106, 107).

222. The dense material of wood is made by the *fixation of carbon*, in combination with other organic elements.

Each complex molecule of wood substance seems to contain within itself 12 atoms of carbon, 8 atoms of oxygen, and 8 atoms of hydrogen. Its composition may therefore be symbolically represented by the accompanying sketch: When wood is made, 12 atoms of carbonic acid (105) are taken, 16 out of their 24 atoms of oxygen are dismissed, and the remaining elements are intimately united with 8 atoms of hydrogen procured from water or other sources, to form one wood molecule. This is then conveyed away, to be packed into some appropriate cell, awaiting its reception elsewhere in the plant, and the superfluous oxygen is rejected through the exhaling pores. Coal is merely wood, whose organic molecules have lost small proportions of water and carbonic acid through the process of decay, until very nearly all the oxygen is removed, and only the carbon and hydrogen remain (103).

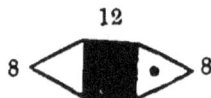

223. Wood substance is, to a *certain extent, liquid*, when it is *first formed* in the leaf.

Hence, it is then easily conveyed back from the leaf, along the channels that lead towards the stem, some of it being, however, absorbed by the way into the several cells that lie near to its course, to form the fibres of the leaf. The fact is, that although the fixation of carbon and the formation of organic material begin in the cells of the leaf, the process is a very slow one, and only completed by the formation of perfect wood in the cells in which the condensed deposit is finally stored away. All the living cells of a plant are transparent, more or less, and therefore carry on the operations of vegetable life in their interiors, altering and condensing the materials that are introduced through their films. The woody material, whose formation commences in the leaf, is conveyed back to the store cells, in which it is at length packed away, through the influence of osmose. The self-same power that causes the rise of the sap into the leaf, determines the flow of the perfected material towards the stem and roots.

224. As the completion and consolidation of wood substance takes place in the *cells of the stem*, oxygen must be disengaged therein as well as in the leaf vesicles.

Oxygen is really conveyed from the stem to the leaf, along channels provided especially for the purpose, and is by them discharged into the loose spaces that are in immediate connection with the exhaling pores.

225. The channels, that are provided for the conveyance of the superfluous oxygen to the leaves, are called *spiral vessels.*

They are so called, because they are tubes lined inside, by means of a dense deposit which is arranged like the spires of a screw. This peculiar construction of the spiral vessel is represented in the figure (*Fig.* 17). Spiral vessels are found in the midst of the fibres of the root, young wood, and leaf, Fig. 17.

and they always contain air with ten per cent. more oxygen in it than the atmospheric fluid does. This oxygen is really the superfluous gas that is set free, wherever the formation of organic material is going on in the living vesicles of the plant, and it is on its way towards the exhaling pores, through which it is to be finally removed. The spiral lining of the vessels is provided as a means of preventing the intrusion of liquid; so soon as the spires get separated or broken, which they commonly do in plants of the fast-growing kind, like the Hop and the Balsam, liquid sap runs in through the cell films, and excludes the air and gas.

226. Living cells from *other complex mat-*

ters out of sap, when it is exposed to sunlight within their transparent spaces, besides wood substance and the structure of their films.

Wood substance and cell film are really the essential ingredients of perfected vegetable structure. Every kind of *structure*, properly so called, that is found in plants, is composed of one or other of these, differently combined and modified to perform diverse offices. There are, however, in the juices, and in the cells of plants, other complex substances produced out of the sap, which are not really possessed of organized structure.

227. These unorganized, but complex products of cell life are destined by nature to be ultimately converted into organized structure: they are, therefore, to be distinguished as ORGANIZABLE BODIES.

The *organizable bodies* are complex products that remain for some time in a sort of intermediate state, before they are finally made into organized textures. They are the organic elements (122) combined together into complex molecules (171), and awaiting the final touch that is to make them the abode of life.

228. *Sugar* is a complex, *organizable body*, produced out of vegetable sap, by the agency of cell life.

Sap is not at all sweet when it enters the rootlets of plants, but it becomes so as it ascends towards the leaves. This is because some of the substances that are comprised within it, are changed into sugar by the influence of the living cells through which they pass. Each complex molecule of sugar is composed of twelve atoms of carbon, twelve atoms of oxygen, and twelve of hydrogen, and its composition may, therefore, be symbolically represented by the accompanying diagram.

12

12 <   > 12

It is formed by twelve atoms of carbonic acid having twelve of their atoms of oxygen driven away, the remaining twelve of oxygen being combined with the twelve of carbon, and with twelve of hydrogen, derived from some other ingredient of the sap. Sugar, therefore, is made by a similar process to that which is employed in the formation of wood substances (222). The only difference is, that the operation of condensation has not been carried quite so far. Less oxygen has been dismissed, and more hydrogen is connected with the other ingredients; therefore, there is *relatively* less carbon in each molecule. Sugar is a less *highly carbonized* substance than wood. The word "sugar" is derived from the Latin term "saccharum," which was taken from an Arabic denomination for a sweet and intoxicating juice (*saccar*).

229. Sugar is formed *for the nourishment* of some of the living textures of plants.

Sugar is found in the cell liquid of nearly all kinds of plants. The sugar that is thus constantly present in vegetables, differs, however, slightly from common *cane sugar*. It has rather more oxygen and hydrogen in its molecules, and is, therefore, a trifle less highly carbonized. It is called *grape sugar*, for distinction's sake. Sugar is very soluble in water, and therefore is particularly well fitted to be easily carried about, wherever nourishment is needed. It is, indeed, the food of all vegetable textures, that do not contain nitrogen in their composition, in its most soluble and active state. The intermediate character of sugar is expressed in its being able to assume the crystalline form, when abstracted from the water which holds it in solution. It is the nature of saline, inorganic compounds, to take the forms of crystals, bounded by straight lines and angles, and of organic compounds and organized bodies to put on the curved outlines of rounded vesicles and grains (8).

230. *Starch* is a complex, *organizable body*, produced out of vegetable sap by the agency of cell life.

Starch is merely sugar a little more highly carbonized. Each of its molecules is composed of 12 atoms of carbon, 10 atoms of oxygen, and 10 atoms of hydrogen; and may be symbolically represented by the accompanying diagram. It is formed, therefore, by 12 atoms of carbonic acid, losing 14 of their atoms of oxygen, and acquiring 10 atoms of hydrogen in their place (228). The word "starch" is derived from the Anglo-Saxon term "*starc*," which signifies "stiff," and was conferred as a name upon this substance, in consequence of the property it possesses of stiffening linen.

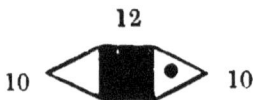

231. Starch is always found *in the interior* of vegetable cells.

Starch is perfectly insoluble in cold water; it, therefore, of necessity remains within the interior of the cell in which it is formed, so long as the cell wall is unbroken.

232. Starch is deposited in the condition of little *rounded grains*, which remain loose and free in the cell liquid.

The molecules of starch are attached together, so that they form little rounded masses. These starch grains are so large that they are readily seen, when sought for by magnifying glasses, but are nevertheless small enough for two or three thousands of them to be capable of lying side by side within the breadth of an inch. The greater part of the granular bodies seen floating in the liquid of living cells, are composed of starch (170). The green color of living vegetables is due to the formation of a thin coating of coloring matter around the starch grains within the cells. Each starch grain, when highly magnified, is found to be covered by circular lines drawn round some one common common point, as *Fig.* 18. represented in the figure (*Fig.* 18). The fact is, that the grain is itself a delicate cell packed full of starch molecules, and these marks are merely puckerings and folds in the film. When starch grains are heated in sulphuric acid, they swell up to three or four times their original size, and the folds are then pulled out by the distension. When starch is placed in boiling water, the cell films are burst, and the starch grains escape, and are changed into a sort of soluble gum.

233. The cells in some plants get *packed completely full* of starch grains, until all other contents are excluded.

In the figure (*Fig.* 19), cells of the potato are *Fig.* 19. shown nearly filled with starch grains. It will be observed that the cells all lie loosely amidst each other; they are in no way attached to the film of the containing cell as woody deposit is. Starch is an extraneous substance, formed in the cell for employment elsewhere, and not an essential portion of vegetable structure.

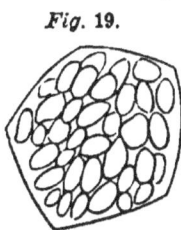

234. Starch is *nourishment* stored away for some future service, *in an insoluble state.*

The starch that is stored away in the potato is designed for the nourishment of the young shoots of the plant when they are produced in the next growing season. When the young shoots of the potato are pushed forth in spring, the tuber is robbed of its starch and loses its "mealy" properties. Starch is deposited in nearly every kind of seed, and, there, it is destined for the support of the young germ that is included, during the earliest stage of its existence. This starch remains fixed and unchanged during the winter, but when the time arrives for the development of the germ, it is converted into soluble sugar, which can be taken up by water, and conveyed into the positions in which its presence is required. The starch deposited in barley, for the nourishment of the contained germ, is prematurely converted into sugar by the process of malting. If the grain had been allowed to fall into the ground, and to remain there until the spring, the rain and sun would have done what the malster effects by his soaking and his stove. When potatoes are boiled that they may be employed as articles of food by man, a somewhat similar change is effected; the starch grains are burst, and the starch is converted into a kind of soluble gum, that is more fitted for immediate digestion. Starch is procured from all parts of plants, but most constantly and abundantly from their seeds.

235. *Oil* is a complex, *organizable body,* produced out of vegetable sap, by the agency of cell life.

Oil is a more highly carbonized product than starch Each of its molecules contains twelve atoms of carbon, one atom of oxygen, and ten atoms of hydrogen. Its composition may therefore be symbolically represented by the annexed diagram. Oil seems to be starch deprived of the greater portion of its oxygen. A starch grain is changed into a globule of oil when it loses nine-tenths of its oxygen.

236. Oil *is found* in small quantities *in most living cells.*

Some of the specks that are seen floating in cell liquid when it is examined by high magnifying powers (170), are composed of very minute globules of oil. There is a very beautiful theory in existence, (which, however, does not yet take rank with the proved discoveries of science,) which considers that young cells are always first formed by the deposit of their thin, organized films around oil globules. This theory assumes that whenever oil drops are placed in connection with the complex substance, out of which cell films are composed, a deposit of this substance is made where the contact with the oil occurs.

237. Oil is stored away *as a fixed deposit* in the cells of many kinds of *seeds*.

Cocoa-nut oil, palm oil, linseed oil, colza (colesecd) oil, and castor oil, are all expressed from the seeds of the plants yielding them. These oils are placed in the seeds to serve as nourishment for the germs contained within them, during the earlier period of their growth.

Fig. 20. The figure (*Fig.* 20) represents a cell of the cocoa-nut packed full of its oil globules. Oil is a very combustible substance, because it, like coal, is almost entirely composed of carbon and hydrogen. When it is burned, these elements are united with oxygen, derived from the air, and carried off as carbonic acid and water (108–70).

238. *Gluten* is a complex, organizable body, produced out of vegetable sap by the agency of cell life.

If any kind of fresh vegetable juice be allowed to stand for a few minutes after it has been pressed from some plant, a greenish-colored, jelly-like substance will be seen to settle to the bottom. The green coloring matter may be easily washed away from this deposit by frequent shakings with water, and subsequent settlings and pourings off; the remainder will then be found to be a greyish-colored body of a sticky or glutinous nature. It has received the name of *gluten,* because this word is the Latin for " glue " or " paste."

239. Gluten is a much *more complex substance* than either wood, sugar, starch or oil.

Those four products of vegetable life are formed without the aid of nitrogen. They are exclusively composed of carbon, hydrogen and oxygen. But gluten contains in itself all the four organic elements (122), and is much nearer in composition to the organized cell films (159). Each molecule of gluten has in itself forty-eight atoms of carbon, thirty-six atoms of hydrogen, six atoms of nitrogen, and fourteen atoms of oxygen, besides a small trace of sulphur and phosphorus. Its composition may therefore be symbolically represented by the accompanying diagram. In other words, gluten consists of 49 per cent. of carbon, 14½ per cent. of nitrogen, 7 per cent. of hydrogen, 14 per cent. of oxygen, and a fraction of a per cent. of sulphur and phosphorus (estimated by weight).* It is made, in the vegetable cell, out of carbonic acid, water and ammonia, by getting rid of superfluous oxygen and hydrogen, and by combining the remaining elements together.

240. Gluten is of a *highly plastic nature*, and peculiarly fitted for the work of organization.

"Plastic" means "adapted for constructive purposes." It is derived from the Greek work "*plastikos*," which signifies "fit for building." Gluten is very prone to decomposition, and is easily caused to decay, but this only makes it more suitable for the purposes of life. Life is a state of change (3) and not of fixedness. Its highest results are produced (as will be hereafter more particularly shown) by rapid change of material. Hence a substance prone to undergo rapid change is the best suited for employment where the higher vital operations are concerned. Gluten owes this useful property of proneness to undergo decay to

---

* Each atom of the different elementary bodies has a different weight. An atom of carbon weighs as much as six atoms of hydrogen; one of oxygen as much as eight atoms of hydrogen; and one of nitrogen as much as fourteen atoms of hydrogen : hence the difference of the numbers that express composition by atoms, and those that express it by weight.

the presence, in its composition, of the inert element, nitrogen (50).

241. Gluten is *sometimes stored away* in vegetable cells in a fixed state, so that a supply of it may be ready for future use.

The fixed store of gluten is chiefly found in seeds. This plastic substance is placed in seeds for the nourishment of the young germs that are to spring thence during their earliest stages of existence. Grains of wheat consist almost entirely of starch and gluten mingled together, in the proportion of nine parts of the former to every two of the latter. If five pounds and a half of wheat flour be washed with successive streams of water while it is kneaded up briskly by the hand, four pounds and a half of starch grains will be taken away in the water, and there will remain one pound of nearly pure gluten sticking as a sort of paste about the hand. When the young germ of the seed is developed, it is probable that the gluten is consumed in building up the living structure; that is, in making the cell films to which its own composition is so nearly allied, while the starch is used in forming wood and the various deposits that are laid upon the films. Hence the distinction of the *organizable products* of vegetable life into *plastic* and *aplastic* principles. The plastic principles contain all the four organic elements in themselves, and are resolved into water, carbonic acid, and ammonia, when they are destroyed. The aplastic principles contain only three of the organic elements, and are resolved into water and carbonic acid alone, when they are destroyed; their destruction, too, is generally of the nature of ordinary burning, hence they are also called *combustible* bodies. Wood, sugar, starch, and oil, may all be resolved into carbonic acid and water, at high temperatures, with the evolution of light and heat.

242. *Albumen* is a complex, *organizable body*, produced out of vegetable sap, by the agency of cell life.

E

If the clear liquid, that remains after gluten has been removed from vegetable juice (238), be boiled for a few minutes, it ceases to be clear, and a white, cloudy-looking deposit appears in it, which slowly subsides, and collects together at the bottom. This deposit is formed of a substance that very closely resembles "white of egg," and that has thence been called "*albumen;*" "albumen" being the Latin name for white of egg.

243. Albumen is a *plastic substance*, like gluten.

Albumen seems to be little more than gluten made soluble in cold water. The two substances are so nearly identical in composition, that the chemist has not been able to find any difference between them. The molecular composition of gluten (239) may be taken to represent also the composition of albumen. There is also another substance, of which the same remark may be made, which is found in abundance in the meal of peas and beans, and which only differs from gluten and albumen in being soluble in both cold and hot water; it is, however, separated from its solution, in a solid clotted state, when an acid is added. This clotted mass is found to be so like to the coagu lum, which separates from milk, when an acid is added in the process of cheese-making, that it is called *caseine,* the word being derived from "*caseum,*" the Latin term for cheese. These three apparently identical productions of vegetable life—gluten, albumen, and caseine— are classed together as *glutinous bodies.*—"Coagulum" means simply a clot; it is taken from the Latin "*coagulo,*" to curdle.

244. Albumen is stored away in a *fixed state* in some seeds.

The white portion of nuts and almonds is almost entirely fixed albumen. The fact seems to be simply this; a complex, plastic principle is made out of vegetable sap, and is stored away round the young germ in the seed, for its future use, but this plastic principle is in the state of *gluten* in some seeds (insoluble in cold water); of albumen in others (soluble in cold water, but not in it when boiling hot); and of caseine in yet others (soluble in both cold and boiling water, but not in dilute acids).

245. Vegetable life is essentially *constructive* in its operations.

All these *organizable products* of vegetable life are complex principles, *built up* out of simpler materials. Carbonic acid, water, and ammonia,—binary compounds,—are the principal food of plants (125); but these are, under the operations of vegetable cells, made into sugar, · starch, oil, wood, gluten, and cell film,—far more rich and complicated principles.

Plants make innumerable other complex substances, beside those enumerated above, upon which the peculiar odors, flavors, and other characters of their textures depend. These however are special, rather than general products. Each different kind of plant yields its own particular substance, and not that which characterizes other species. One forms an essential and aromatic oil, another a bitter extract, and another an astringent principle; but all alike yield sugar, starch, fixed oil, and gluten. The remark made above, applies, nevertheless, to these special productions with equal force; they are all complex substances *constructed* out of simpler materials.

246. The constructive operations of vegetable life are mainly performed by *fixing carbon, hydrogen, and nitrogen*, and by *dismissing oxygen*.

This is an invariable rule in the vegetable economy, and is illustrated in the production of all the organizable substances described above. Whenever living, vegetable cells are at their constructive work, streams of pure oxygen are poured off from them into the air. Vegeta-

tion is constantly restoring to the air oxygen that is removed by the processes of combustion, and of oxidation
in its countless diversities of form (35).   Vegetable life
is thus the grand antagonist of mere chemical power.   It
is always doing what the simple operations of chemistry
are altogether unable to effect (106–107).

247. Living vegetables are enabled to effect
their office of *constructively fixing carbon,
hydrogen, and nitrogen, and of exhaling oxygen,* through the stimulant influence of solar
light (6).

Hence, it is, that leaves, which are the most energetic
and vital portions of plants, are so carefully spread out
into the atmosphere, and towards the sky.   Whenever
growing plants are placed in darkness, they cease to
construct complex products, and to exhale oxygen, and
they pour forth streams of unchanged carbonic acid instead.   Plants which are kept in the dark, form no
wood, or starch, or coloring matter; they send out pale,
straggling, weak shoots, and unfold white leaves, in the
place of their usual verdant organs.   Celery is artificially blanched for the table, by being reared in the absence of light.   All the operations of vegetable life go
on with the utmost vigor during the season of summer,
because the solar light is at that time very intense; and
they are languid or even dormant through the winter,
because then the solar influence is comparatively small.

CHAPTER X.

THE OBJECT OF ANIMAL ORGANIZATION.

248. Animal bodies are made of *different kinds of organs*, so contrived and adjusted with regard to each other, that they all contribute to one common end (2).

Animals have limbs which carry them about, stomachs which digest their food, hearts which circulate their blood, and various other distinct organs that are fitted for the performance of other particular offices. Plants, too, are composed of distinct organs, but, in them, these organs are all like to each other. Each leaf in a tree resembles the rest. Even flowers have been shown to be only clusters of leaves slightly modified, to enable them to perfect the seed. The largest trees can do nothing more than the smallest leaf on one of their twigs can perform unaided and alone. An animal, on the other hand, can accomplish a host of things, which no separate organ in its frame could effect. Yet every separate organ contributes something that is necessary to the completeness of the result. A leaf bud may be taken away from a tree, without causing the tree to suffer any inconvenience, but a stomach, a heart, or a limb cannot be taken away from an animal, without destroying, or maiming the completeness of its body. Different organs, each capable of performing a different office, are conferred upon animals, in order that the animal may avail itself of the service of the whole, and through them accomplish a wide and varied series of operations, far exceeding in importance those that the most perfect plants can effect. The completeness is in each separate organ in the plant; but in the animal, it is in the body of the creature taken as a whole (198).

249. The organs of animal bodies are *built up*, principally, *of cells.*

There are very few portions of the animal body, in which traces of cell formation may not be discovered, by the skilful employment of the microscope. The organs of animals are built up of cells, as houses are built of bricks (1).

250. The simplest animals are formed of *single, isolated cells.*

Some of the unicellular organisms (182) are unquestionably of an animal nature. *One-celled animals* are generally found floating about in stagnant water. They are of very minute size, and can only be discerned by the aid of the microscope; hence they belong to the tribe of what scientific men have agreed to call *animalcules,* or "*little animals.*" The smallest animal known is a minute cell that inhabits stagnant water, and that has been named the "Twilight Monad," (*Monas crepusculum* : "*Monas—Greek for "unity,*") it being the unit that forms the dawn, so to speak, of animal existence. This monad is of so slight a stature that a   *Fig. 21.* single small drop of water could accommodate eight thousand million individuals of its tribe.

The figure (*Fig.* 21) is a small group of portraits of the Monad magnified more than half a million of times.

251. The one-celled animals *possess much more power of motion* than the one-celled plants.

The Monad may be watched swimming about in a drop of water, with extreme vivacity. The Diatoms, (once taken to be of an animal nature, but now known to be really plants,) on the other hand, scarcely ever move, and when they do so, scarcely get through a journey of an inch in an hour. The Monad rows itself about adroitly, by means of a delicate, elastic oar, called a *cilium* (from the Latin word for hair), which it waves rapidly backwards and forwards. This oar cannot be distinguished in the "Twilight Monad," but it may be easily seen in some nearly allied species of larger size. The Diatom, when it moves, merely shifts its position by a series of jerks along a straight line, and if it chances to knock its end against any opposition, lies there helplessly without the power of altering the direction of its course. The movements of the Diatoms are most probably accidental movements, resulting from the impulse of currents of water, set up by the osmose action of their own cell films, rather than true voluntary acts of the creature.

252. Animal cells feed on *complex, organic matters*, and not on simple, inorganic principles, such as serve for the support of vegetable cells (125).

This is illustrated by the difference of the habits of the Diatoms, and the true animalcules. The Diatoms dwell by preference in clear water that can just supply them with a little carbonic acid and ammonia, and with light enough to stimulate their films to fix the carbon, and reject the oxygen. Animalcules, on the other hand, abound in water that is turbid and impure, from the presence of decomposing organic substances.

253. Animal life is *destructive* in its operations.

Animal cells destroy the organic matters they feed on. They convert portions of that material into their own structure first, but then that structure is immediately afterwards decomposed into its elements. Oxygen is taken from the air, and caused to unite with the carbon and hydrogen, and to fly off with them as carbonic acid and water, and the nitrogen escapes with other portions of the hydrogen as ammonia. Animal life is thus exactly

antagonistic of vegetable life, so far as its operations are
concerned (245, 246). Animals undo what plants have
done.

254. Animal cells always *imbibe oxygen*,
and *exhale carbonic acid* and ammonia, so
long as they continue alive.

The carbonic acid and ammonia exhaled are the products of the destruction of organic matter employed as
food. That which has been made out of carbonic acid
and ammonia, by the life of the plant, is turned again
into carbonic acid and ammonia to support the life of the
animal. Vegetable and animal organization meet so
nearly in the unicellular organisms that it is sometimes
a very difficult task to determine to which kingdom certain living forms, under notice, belong. The only way
in which a safe conclusion can be reached, under these
circumstances, is by ascertaining the nature of the operations the doubtful creature performs. If it lives on
carbonic acid and ammonia, converts them into complex,
organic substance, and exhales oxygen, it is unquestionably a vegetable; if it lives on complex organic matter,
imbibes oxygen, and exhales carbonic acid and ammonia,
it is no less certainly of an animal nature.

255. Animal life *restores to the air* that
which vegetable life consumes.

Living animals exhale carbonic acid and ammonia,
which are mingled with the atmosphere; but carbonic
acid and ammonia are the essential principles that are
conveyed from the atmosphere and soil into the plant,
by the winds and rain, to serve it as food (125).

256. Animal life *takes from the air* that
which vegetable life supplies.

The operations of animal life are dependent upon the
presence of pure oxygen (as will be presently seen,)
which is continually united to the elements of the animated structures in the formation of carbonic acid and

ammonia ; but the operations of vegetable life are constantly restoring pure oxygen to the air. (217–218.)

257. Animals and plants are thus *compensatory*, as well as *antagonistic* (253) in their operations.

What is waste to plants is essential to the vital operations of animals. What is waste to animals is the ordinary food of plants, and (as will be hereafter seen) plants furnish a continued supply of food for animals, at the same time that animals furnish the supply of food for plants.

258. Animal life is destructive, because it gets *power out of the material* it destroys.

No effort can be made by any animal organ without some part of the organization being destroyed. A man cannot lift his hand to his head without some portion of the muscle of his arm being wasted in the production of the movement. He cannot use his eye, nor his ear, in observing the appearance of an object, or in listening to the tone of a sound, without dissolving some nerve-molecules into their elements. The great object of animal life is the production of force, out of the consumption of material substance. Motion is as much *got out of* muscular flesh, at the expense of its consumption, as heat is got out of coal, by its combustion, when it is burned.

259. Animal cells perform all *the ordinary offices of cell life* (180); excepting the production of organic substance out of inorganic elements.

It is in this one respect, that the *destructive* animal cell differs from the *constructive* vegetable cell. All the other four functions of cell life may be seen to be performed by the single-celled animalcules. They grow by attaching the organic substances, on which they feed, to their own textures. They vitalize the substance they

E*

attach to their textures, and enable it to perform work like their own. They alter their forms as they advance towards maturity, and they produce offspring like in all respects to themselves, by subdividing and budding. These several organic offices of animal life may be watched, proceeding in their simplest and least complicated conditions, in the microscopic animalcules that swarm in stagnant water.

260. The film of the animal cell is generally very *mobile and limp.*

The animal cell is mostly distinguishable from the vegetable cell in this way. It tumbles about easily in all directions, like a bladder filled with water, and readily assumes any form; while the vegetable cell is more or less rigid, and preserves one unvarying outline. This is well illustrated in many of the animalcule tribe, whose little vesicles may be observed altering their general appearance at every instant, as they row themselves about, but the character is the most beautifully shown in one species, which is the largest of all the known one-celled animals. This creature inhabits impure water, in which decaying organic matter abounds, and is large enough to be just detected by the naked eye; its dimensions are, in fact, so magnificent for an animalcule, that not more than a thousand of its portly vesicles could be laid within the space marked out by the line in paragraph 206. It is called the Amœba, or the "*changeable*" (*from "Ameibo," the Greek word for "to change.")* When at rest, it looks like a small round bag of soft, filmy membrane. The drawing *a* (*Fig.* 22) shows its appearance when magnified something more than a thousand times. This little bag, however, is alive, and needs to be fed. Its proper nourishment is the morsels of decaying organic substance that float in the liquid in which it dwells; these morsels, therefore, it seeks out for itself, but it accomplishes this purpose in a most ex-

a   Fig. 22.   b

traordinary fashion. It pushes out one side of its film somewhat in the form of a finger of a glove, then rolls over its entire, limp body, after the finger, and next projects another finger in some other direction, until it has assumed some such grotesque and indescribable outline as that represented at *b* in *Fig.* 22. If, during the progress of this series of changes, the cell film chances to come somewhere in contact with a fragment of food that is to the creature's taste, it quickly folds itself over the morsel, until this is entirely enclosed in a sort of pouch, or extemporaneous stomach, formed especially for its digestion. Here the fragment lies, in close contact with the enveloping film, until it is dissolved and absorbed by osmose into the cavity of the cell. Then the fold relaxes itself, and opens out to be ready for the next prey it may light upon. The Amœba does not seem to be possessed of any consciousness, or will. It is merely a limp, living bag, rolled over this way and that, in all possible directions, in the certainty that where decaying matters are so abundant, some appropriate food must soon be entangled within its folds.

261. Every animal *begins its existence*, as a single, living cell.

Man himself, in the earliest stage of his existence, is merely a small vesicle, possessing the power of attaching to itself organic substance, and of producing a succession of similar vesicles through subsequent generations.

262. In the highly developed animal, the successive generations of cells, produced from the parent vesicle, *are connected together* to form the various textures and organs.

The cells, out of which the organs of plants are built, are all successive generations of individuals, produced by the ordinary processes of cell division and cell budding from one parent vesicle and its descendants. The chief difference between the one-celled animal organism, and the perfect animal, is, that the cell offspring of the former are scattered loose into independent existence, as soon as they are formed, while the cell offspring of the latter are retained for a certain time within the body, connected together, and in this state made to do service as organs (185).

263. The bodies of animals *increase in size* only until they reach a certain fixed standard of growth.

Oak trees sometimes attain to an enormous size. Their dimensions are only limited by the difficulty of furnishing nourishment to the foliage, when this is stretched very far from the roots. Elephants and whales, on the other hand, speedily attain a fixed, mature size, and then do not grow any larger although they continue to live for years afterwards.

264. In the mature animal, *cells are destroyed* as fast as they are formed.

All the operations of animal life are carried on by the generation and destruction of cells (258). The new cells formed in the plant, excepting the leaf cells, are attached permanently to the stem to increase its bulk. But the new cells of the animal are dismissed from the body as waste material, so soon as they have contributed to the performance of certain vital operations. Hence it is that animals do not grow beyond a given standard of size, although they are engaged in the production of new cells all their lives long.

265. The substance of which organs are formed is continually undergoing *removal and renewal.*

The vital powers of organs result from this change of substance (258). Life is a state of change (3).

266. The *duration of animal life* depends on the continuance of cell production.

So long as new broods of cells are produced, all the

various organic changes, which collectively constitute
the life of the individual, are carried on.

267. Death results from the *exhaustion of the power of cell multiplication.*

The primitive vesicle of every animal body hands
down its own reproductive power to the successive
generations of cells, that are called into existence as its
offspring. This power is transmitted through countless
myriads of generations. Millions upon millions of
cells, in inconceivable numbers, are produced from the
one parent. Still there at last comes a time when this
reproductive power is exhausted by its own operations.
Then, no more cells are produced. All the several
organs are consumed in their proper operations, being
no longer renewed as they were before. Consequently
their activity stops, and death takes the place of organic
life.

268. Some cells are retained in a *condition of comparative permanence* in the animal body.

But such cells are then no longer living bodies.
They are killed and packed full of dense deposit, which
ensures their preservation, very much as the dense
deposit of the wood cells of plants does. Bones are
constructed in this way. All such dense, dead and permanent cells are, however, only used for mechanical
purposes in the animal frame (191); this is instanced
in the bones, which merely form a frame-work for the
more important parts to be distributed round. They
are never employed in vital offices. Living organs are
invariably made of cells that have but a very transient
existence, and that are constantly being reproduced and
destroyed (189); and, as a general rule, the most active
organs are formed of cells, that have the shortest existence, and are changed the most rapidly. This must be
the case, since the power of animal life is got out of the
destruction of organic substance (258).

269. Animals *require food* for the production of the successive generations of their cells.

The food of animals supplies the organic material that is first built into the structures of the various organs, and that is then destroyed in the production of power. The food of plants furnishes materials for their constructive work, and gets stored away as accumulating organic substance. The food of animals furnishes them with vital power, and is dissipated in waste as fast as their organs are employed.

270. The cells of the animal body are *rendered perfect for the work* they have to do, subsequently to their first formation.

All cells in the animal body are at first simple, soft, film vesicles, like the parent *out* of which they have descended. But some then get so altered in their character as to fit them to make bone. Others become suited to make gristle, others tendon, others muscle, others blood-vessels, and others skin. It is in this way that *different kinds of organs* are built up in the animal frame, out of simple cells.

271. Animal organization is subservient to the support of *animal life.*

Animals possess something beyond *organic life ;* that is, the capability of effecting changes by cell activity— producing successive broods of active cells, and combining them into special organs (259). This superadded something, existing in the organic frame of animals, it is, that is known physiologically by the name of *animal life.*

272. Animal life consists of the power of *moving at will,* and the capability of *perceiving* the presence of external objects.

Plants have no power of altering the positions in which they stand, or of adapting themselves to circumstances. Animals, on the other hand, are expressly constructed, so that they may acquire information regarding the objects around them, and so that they may be able to control and determine the relations in which they are to stand to those objects. A strawberry plant does not move when a heavy foot hangs over, just about to press down upon and crush it; but a frog, in the same position, sees the foot—understands that there is danger in it to its own organization—puts its jointed limbs to work—and hops out of the way. This ability to *move at will*, and to *perceive* extraneous bodies and relations, is the especial privilege of animal life; it is, indeed, the power which animals get out of the destruction of organic material (258). The entire organization of the animal frame is planned with a view to the production of this result. The organic life of the perfect animal is entirely subservient to its motor and perceptive endowment. All the various organs of its body work, merely that it may be sustained in its capability to move about, and to vary its relations to extraneous things.

CHAPTER XI.

FOOD AND DIGESTION.

273. The activity of animal bodies is supported by the *repeated taking of food*.

Since every operation of the animal body is attended with waste of its substance (253), it is clear, fresh supplies of material must be furnished from time to time, if the frame is to be maintained in a working state. These supplies of material are furnished as food.

274. Animals employ as food the *organizable substances* that are formed by the constructive operations of plants (227).

Plants form these substances as a store of nourishment to be employed in the support of their fresh shoots, or in the development of the new individuals contained within their seeds, when the season of growth returns. Hence the "organizable products" of vegetable life are deposited either in the stem, near to where the new shoots will be pushed forth, (potatoes are underground stems) or in the seeds, around the latent germs. Animals, however, seize upon this store, and appropriate it to their own uses. They feed upon that which the plant has providently accumulated. Flesh-feeders do not, in reality, make any exception to the conditions of this statement, for the animal substances that they consume have been primarily nourished by vegetable material. Man eats beef, but that beef has been extracted from the grass of the field.

275. Every kind of substance that is employed by the animal as food consists of *sugar, starch, oil, or glutinous matters,* more or less mingled together in various proportions.

Ripe fruits consist chiefly of solutions of sugar. Succulent roots, stems, and leaves contain starch grains in their cells, and gluten and albumen in their juices. Potatoes and rice are starch, with a small admixture of gluten. Wheat flour is starch and gluten, in the proportions of four and a half parts of the former to one of the latter. Peas and beans are starch and caseine. Almonds and many other kinds of seeds are almost exclusively albumen. Most seeds contain oil. Flesh is fat (oil) and gluten in a slightly altered state.

276. Sugar, starch, oil, and glutinous matters are the *nutritious principles* that have

been designed for the support of the animal frame.

The destructive animal consumes what the constructive plant forms (257), and, therefore, gets its power out of the results of vegetable activity. In the design of nature, the vegetable economy has been made constructive, in order that the animal economy may be endowed with its especial privileges (272).

277. The glutinous principles of food are employed in *building up* the animal structures.

Gluten—albumen—caseine (and flesh) are all plastic (240). They are fitted to become animal structure, in consequence of their being composed of all the four organized elements (239). The presence of nitrogen disposes them to suffer that ready decomposition, which the necessities of animal life require.

278. The *oil, starch, and sugar* of food are *chiefly combustible* principles. They are the fuel which is used in the heating service of the body.

In order that the several changes, upon which the maintenance of animal life and power depend, may go on actively, it is necessary that the frame should be maintained at a warm temperature. The body is kept nearly forty degrees hotter than the mean temperature of the air in which it is placed, even in warm climates. This result is attained by the slow burning of fuel in its interior. The fuel that is so burned is the oil, starch, and sugar of the food. It will be remembered that these are all principally composed of the two combustible elements, carbon and hydrogen (228–230–235), and are, therefore, themselves of a highly combustible nature. The provisions that are made for burning fuel within the animal body, and so maintaining its elevated temperature, will be hereafter explained.

279. The nutritive principles of the food consumed by animals, are, in the main, *solid and insoluble bodies.*

It has been seen that they are mostly fixed stores, packed away by plants for future use : they must be in the solid state and insoluble to answer this purpose, as otherwise they would be continually carried away in the vegetable juices that are able to permeate the cells. Starch is always in the condition of little solid grains, contained within cells (232) ; and the glutinous principles of seeds are always dense masses (as instanced in the firmness of corn grains, and of beans and peas). Oil is insoluble, but it is itself a liquid, without the aid of water ; it is, however, always packed away in cell films, while held in vegetable structures. Sugar is the only one of the nutritious principles that is really soluble ; hence it is never found in a fixed state, unless when fruits, whose juices contain it, have been artificially dried.

280. Nourishment cannot be made serviceable for the support of animal structures, unless it is *in the liquid state.*

The nourishing material has to be conveyed into the interior of the living cells, of which animal structures are composed (249), otherwise it could not be changed into living substance under their influence. In order that it may be able to permeate the films of these cells, it must, therefore, be a thin liquid, like water. The nourishment of the animal frame is chiefly in a solid state, when it is taken as food ; but it is in the liquid state when it is applied to the renewal of exhausted structure.

281. The solid principles of the food are rendered liquid by the process of *digestion.*

The liquefaction of the food is effected by means of an apparatus furnished to the animal for this especial purpose. This is called the *digestive apparatus*, and consists of a series of separate organs, performing different offices. The word digestion merely means solution. It is derived from the Latin word " *digero*," to "divide" or " dissolve."

282. The first step in the digestive process is the *breaking up* of the substances, used as food, into fragments.

This operation is performed in the mouth by means of jaws and teeth. The food is turned about in the mouth by the tongue, and is cut, rasped, and ground by pressing it between the teeth, and then moving the teeth backwards and forwards upon each other, until it is reduced into a state of very fine division. The act of grinding the food in the mouth, is called mastication or chewing. " *Masticatio*" is the Latin term for chewing.

283. The food is *mingled with saliva* as it is crushed between the teeth.

In this way, the ground food is converted into a sort of paste. The saliva is a viscid liquid, that is formed by little organs, placed beneath the tongue, and at the back of the lower jaw; it is poured out into the mouth, through small tubes that are pierced through the lining of that cavity. "Saliva" is derived from the Latin word *Sal*, " salt." It is so called because it possesses a saline taste, which is due to the presence in it of a small quantity of alkaline and earthy salts. The " tartar" which is apt to accumulate on the teeth, consists principally of deposits of these salts, derived from the saliva. The saliva is pressed out of the salivary tubes into the mouth by the movements of the jaws. It plays a very important part in the operations of digestion; nearly a pint of it is furnished every twenty-four hours for the use of an adult man.

284. The saliva renders the *starch of the food* soluble, and makes its other principles more ready to mingle with water.

These changes commence in the mouth, and go on

steadily after the food has been swallowed. When
starch is ground fine and mixed with saliva, it is slowly
and gradually changed into a sort of sugar.

**285. When the food has been ground
fine, and well mixed with saliva, it is *swallowed.***

The food and saliva, mingled together, are forced down
the swallow into a sort of pouch or bag, contained within
the interior of the body, and formed of strong membrane. This bag, prepared for the reception of the masticated food, is called the *stomach.* The word is derived
from the Greek terms *stoma*, the mouth, and *cheo*, to
pour. It is the cavity into which the contents of the
mouth are poured.

**286. In the stomach an *acid liquid* is
added to the masticated food and saliva.**

This stomach liquid is called the *gastric* juice. The
word gastric is taken from the Greek word "*gaster*,"

Fig. 28.

which signifies "stomach." The gastric juice is furnished by a series of
delicate tubes which lie in bundles
within the walls of the stomach, and
open into its cavity by a myriad of
mouths. In the figure (*Fig.* 23) a
magnified view is given of a portion
of the inner lining of the stomach,
showing the mouths of the tubes that yield the gastric
juice, connected together in bundles.

**287. Gastric juice is poured into the stomach
whenever *food is received* into the organ.**

Immediately after taking a meal, gastric juice is added
in considerable quantity to the masticated food that has
been swallowed. The mere presence of masticated food
in the stomach stimulates its tubes, until they send forth
the gastric juice. The whole contents of the bag are
then intimately mixed together, by a sort of churning
motion of its walls, which is repeated from time to time.

288. The gastric juice consists of a *dilute acid*, mingled with a peculiar organic ferment containing nitrogen.

The acid contained in the stomach is that which is known to the chemist under the name of *hydrochloric acid*. It is one of the compounds in which the element chlorine takes the ordinary place of oxygen (148). Hydrochloric acid is salt in which the sodium is replaced by hydrogen (70). It is composed of chlorine and hydrogen united together in a certain definite proportion. The ferment of the gastric juice is a decomposing substance, something of the nature of yeast. It is called *pepsin*, a word derived from the Greek "*pepto*," "to digest."

289. The *glutinous ingredients* of the food are dissolved by the gastric juice.

The same result is effected out of the stomach, by boiling in hydrochloric acid; the digestion of the glutinous principles by the gastric juice, is, therefore, a simple chemical operation. A boiling heat is not required for the process in the stomach, because the organic ferment (pepsin) disposes these principles to be dissolved by a weak acid, at the ordinary temperature of the body, or even in greater degrees of cold. This predisposition to be easily soluble in acid it is the office of the pepsin of the gastric juice to effect. That the gastric juice acts as a simple chemical solvent, when it performs its digestive duty, is proved by the fact that it digests the walls of the stomach itself, when left in contact with them after death.

290. After being well mingled with the gastric juice, and after being allowed to remain in communication with it in the stomach, for some time, the digesting mass *is pushed forwards into the bowel*.

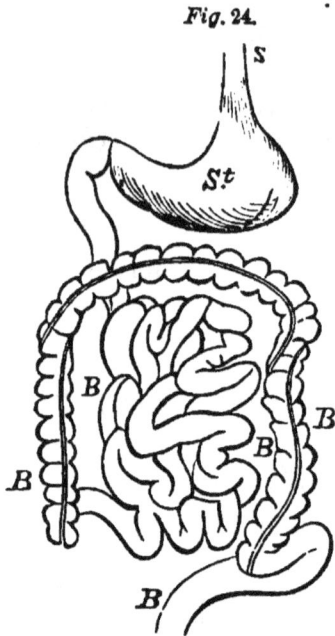

Fig. 24.

The further extremity of the stomach is narrowed and prolonged into a tube, which is many feet long, and coiled up in folds for the convenience of stowage. This coiled-up tube is called the *Bowel.* (The word is a corruption of the old French term for the same organ— "*Boyau.*") The general form and arrangement of the stomach and bowel, constituting together what is called the digestive canal, is shown in the accompanying figure, (Fig. 24). *S* is the swallow coming down from the mouth. *St* is the stomach, and *BB* are the folds of the bowel.

291. In the bowel, the half-digested food is mixed with *two other fluids*, that are poured out from ducts which open into the canal.

These bowel contributions to the digestive process are called *the Bile* and the *Pancreatic Juice.* The word Bile is derived from the Latin term " *Bilis,*" " choler," or " anger." The ancients fancied that the substance, to which they gave this name, had something to do with the production of fierce and ill humors in men. The word *Pancreas* is derived from two Greek words *pas* and *kreas*, which signify " all flesh ;" it was first employed on account of the dense, fleshy nature of the organ from which the juice is derived. The Bile is formed by a similar large organ called the *Liver*, which is placed just above the stomach. (*Liver* is taken from " *Lyfer*," the Anglo-Saxon name by which the organ was called.)

292. The bile and pancreatic juice enable *the digestion of the starch of the food to be completed.*

The bile and pancreatic juice are of an alkaline nature; they contain free soda, amongst other things. They neutralize and render inert the acid of the gastric juice, when they are mixed with the digesting substances in the bowel, but this is not done until all the glutinous matters have been thoroughly dissolved. So long as the acid gastric juice is present, the digestion of the starch, by the saliva, is arrested; when, however, the gastric juice has fulfilled its offices upon the glutinous matters, and its sharpness is blunted by the bile and pancreatic juice, the saliva resumes its work, and pursues it to completion.

293. The bile and pancreatic juice also render the *oily ingredients of the food* fit for admission into the cells of the system.

They do this by converting them into a kind of *soap*, which is, to a certain extent, soluble in water. Soap is made by combining oil and alkaline substances together. The free alkali of the bile unites with the oil of the food, and so changes it into a soapy substance.

294. The bile and the pancreatic juice also *separate the digested principles* of the food from the coarse undigested refuse.

After the bile and pancreatic juice have been for some time in communication with the food in the bowel, *digestion is complete;* that is, all the nutritious principles contained in the food are reduced into a liquid, or soluble state, so that they can be taken up by water, and can be conveyed by its agency through delicate films and tubes. The saliva, gastric juice, bile, and pancreatic juice, together form a combination of solvents, which none of the organizable products contained within the food are able to resist. By their agency one after the

other is reduced from its fixed condition, and so made
amenable to the transporting powers of water.

**295. The digested nourishment is called
chyle.**

" *Chyle*" is derived from the Greek word " *chilos*,"
which signifies food. Chyle is a white liquid, very
much resembling milk in appearance. It is merely oil,
sugar, and liquefied starch and glutinous matters, all
mixed up together into a sort of *emulsion*, or " milk."

**296. The digested, or liquefied food (*chyle*)
is *sucked up* from the bowel into a series of
small vessels, provided for effecting its absorption.**

The inner coat of the bowel is covered by an immense
number of little projecting points, which hang down
into the cavity of the canal, like the soft pile of velvet :
these are very minute. They are about the thirtieth of
an inch long, and some four or five hundred of them are
contained within a space as large as that enclosed
between the lines of the annexed diagram, so
that there are some millions of them altogether
in the bowel. They are called *villi ;* a
word derived from *villus*, the Latin term
for " *wool*," or " *nap*." Each villus has
a cluster of delicate cells arranged about
its point, and a series of vessels running
in from these. This structure is shown
in the accompanying sketch (*Fig.* 25).
When the chyle comes into contact with
the cells, it is absorbed into their interiors
under the influence of osmose (163), and
is thence pushed on into the vessels beyond, by the same power.

Fig. 25.

**297. The coarser portions of the food,
which the saliva and gastric juice cannot dissolve, *are rejected* from the bowel, when the**

nutritious chyle has been removed from them.

The nutritious principles, which can be dissolved by the saliva and gastric juice, are mixed up in the food with coarser substances, really the remains of the fibres and cells in which those principles have been stored away. The chyle, or digested food is separated from this undigested mass, through the influence of the bile and pancreatic juice, and is then gradually absorbed by the villi of the bowel, until the whole is removed. This is effected as the digested mass is pushed gently along by the alternating movements of the bowel. At length the undigested fibrous portion is left, devoid of any admixture with nourishing material, and is therefore rejected from the digestive canal as of no further use. Digestion is thus a separation of the nutritive principles of the food from coarser and useless matters, as well as a reduction of them into the soluble condition.

298. The digestive canal is, in a physiological sense, *an external passage*.

The digestive canal, although contained within the body, is really external to the living system. It is open to the external air, and is lined by a membrane, which is merely a prolongation of the external covering of the body,—the skin,—altered to suit the new circumstances in which it is placed. In some of the simpler forms of animal existence, the body may be turned inside out without the vital actions of the creature suffering any great amount of disturbance: the skin becomes stomach, and the stomach skin, and all then proceeds as before the inversion. The stomach seems, after all, mainly an adaptation to the habits of locomotion for which the animated creature is designed. Bodies, that are to spend their lives wandering about from place to place, cannot send rootlets down into some fixed reservoir of food, as plants do; hence, they have bags within their frames, in which they can carry supplies of nourishment about with them, and their absorbing rootlets are

r

distributed to these bags. The villi of the intestines are really rootlets of absorption, as much as any spongiole that sucks in sap from the soil.

299. The apparatus of digestion is chiefly packed away into one large cavity of the body.

The cavity of the body, in which the organs of digestion are placed, is called the *Abdomen*. The word is derived from the Latin "*abdo*" "to hide." It is so termed, because the organs, which it contains, are hidden away in it from sight. The abdomen is, then, the especial province of digestion. In it, all the processes of the operation, (saving chewing, and the formation of saliva) are carried on. The lower part of the human body is mainly composed of the abdomen and its organs. The backbone and its muscles are behind the cavity; the front and sides are formed by walls of muscle and skin. Its floor is a basin-shaped bone, which is supported on the lower limbs, and which constitutes the foundation of the bony frame. Its ceiling is an arched, muscular dome, which divides it from the upper cavity of the body, the chest. In the accompanying figure (*Fig.* 26), the relative position of the various organs of digestion is sketched, as they lie in the abdomen. The whole are supposed to be seen in profile, the side of the body being towards the observer. *S* represents the stomach; *BB* the folds of the bowel; *L* the liver; *P* the pancreas; *D* is the position of the arched ceiling of the abdomen, which separates it from the cavity of the chest above.

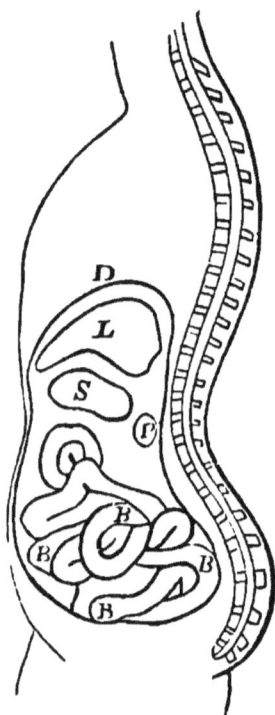

Fig. 26.

CHAPTER XII.

THE DIET OF MAN.

300. Food is taken by man to *support the waste* of his body, and to *keep up its heat.*

Every vital action, that is performed in the animated frame, is effected at the expense of some portion of its structure. It is wasted by its own actions, as surely as the moving machine is worn away by its own operations. Hence, new structure has to be built up as fast as the old is destroyed, and the material of this new structure is furnished from time to time, as it is required, by the introduction of food. But, in order that the vital actions of the human body may be effectually carried on, it is necessary that its organs shall be kept in a temperature that is warmer than the atmosphere. This warm temperature is provided by a constant, slow consumption of fuel within the frame. The fuel so consumed is furnished from time to time by the food.

301. The different organizable principles, contained in the food, possess *different powers* of nourishing and warming.

Some kinds of food, therefore, *nourish* more than others ; or, in other words, repair the waste of the organs more promptly. Other kinds *heat* more.

302. The *glutinous* principles contained in food are the *most nutritious.*

Gluten, albumen, caseine, and lean meat, are all classed together as glutinous principles, in consequence of their being nearly of the same nature (243). They all contain nitrogen, are of a highly plastic character (240), and are converted into organized structure in the living body.

303. The *oily* principles contained in food, are the *most heating*.

They are so, because they are almost exclusively composed of carbon and hydrogen (235), which are both very combustible elements, and convertible into carbonic acid and water, with the production of free heat, by union with oxygen (47). The oily principles of the food are chiefly consumed in the system as fuel, and in the heating service of the body, without having first been made into organized structure.

304. The oily principles of the food are, however, *nutritious in a degree*.

The illustrious German chemist, Liebig, asserts that the plastic nitrogenized principles of the food are exclusively nutritive, and the other principles exclusively combustive, and employed in the heating service. There is no doubt that, in the main, these diverse principles are thus differently used; but it is also now known that some heat is set free when the plastic principles of the structures are decomposed, although very much less in amount than that which is procured from the burning of the same quantity of oil; and so, again, that oil is used in some constructive purposes, although generally of much less service *plastically* than the nitrogenized compounds. The heating power of the plastic principles is instanced in the case of carnivorous animals, which keep up a high temperature within their frames, when they get nothing but lean flesh for their food. The constructive power of oil is shown in the influence of cod-liver oil in restoring the wasted flesh of consumptive people. The fact, therefore, is, as stated above, that the glutinous or plastic principles of food are the *most nutritious*, and the oily principles the *most heating*.

305. The *farinaceous and saccharine* principles of the food are *chiefly heating*, although in a less degree than oil.

"*Farina*" is the Latin name for "starch," and "*Saccharum*" the Latin name for sugar. Hence, starch and sugar are termed farinaceous and saccharine principles.

Sugar and starch are less combustible than oil, because there is already some oxygen combined with their combustible elements—carbon and hydrogen (228–230). They are converted into oil in the system, before they are burned, as will be hereafter seen.

306. The glutinous, oleaginous, and farinaceous principles are *all necessary* for the sustenance of life.

When animals are fed for a long time on any one of these principles alone, they die of starvation, just as they would if they were kept from food altogether.

307. The best diet is that which *combines these several principles* in itself, in the proportion in which they are required by the system.

In most of the substances employed as food, these several principles are naturally combined together, but some foods have more of the one than of the other. Hence, some foods are more nutritious, and others more heating, according to their constitution. "*Diet*" means food, viewed in relation to its character and effects on the living body. The word is derived from the Greek term "*diaita*," which signifies "*rule of living.*"

308. All the several principles are contained *in milk*, in the relative proportions required for the formation of a perfect food.

Milk is nature's own compound, prepared for the

support of young animals, before they undertake the
business of foraging for themselves. It contains all the
principles that are needed for nourishing and warming
their frames. It has in itself glutinous matter, oil, and
sugar, (the representative of the farinaceous principle).
The glutinous principle is caseine (243), which is separated from the other ingredients in the process of *cheese-
making*. The oil is churned out of it as *butter*. The
sugar is held in solution in the *whey*, and is at once
detected there by the sweetness of the taste. One
hundred ounces of cow's milk consist of ten ounces of
solid matter held diffused in ninety ounces of water.
Of the ten ounces of solid matter, about two and a half
are glutinous matter (cheese), and seven and a half
butter and sugar (oleaginous and saccharine principles).
Human milk contains a larger proportion of oil: there
are in it two ounces only of cheese (caseine) to every
eight ounces of butter. From this it may be inferred
that four to one are about the proportions in which the
combustible and the plastic constituents need to be
mingled in the food of man.

309. Wheat flour contains *glutinous and
farinaceous* principles, mingled together in
very nearly the same proportions that similar
principles are in milk.

Wheat flour consists of starch and gluten mixed
together, in the proportions of four and a half of the
former, to one of the latter. It differs from milk
chiefly in the solid and crude condition in which these
stored-up constituents are, and in the absence of water.
When water is added, it is sufficient for the support of
human life for considerable periods of time. Hence,
bread is in such general use among mankind, and hence,
in strict accuracy, it deserves the name that is commonly given to it, of " The Staff of Life."

310. Many of the common admixtures of
different kinds of food are judicious, on

account of their *mingling together opposite principles.*

Thus butter is added to bread, because in flour the combustible principle is in the fixed state of starch, which requires to be converted into oil before it can become serviceable. Potatoes and rice (which are principally composed of starch) are eaten with meat, because it is deficient in farinaceous matter. The addition of butter to bread augments its heat-sustaining powers. The addition of meat to potatoes increases their nutritive capacities. A mixed diet of bread, butter, meat, and potatoes, furnishes all the ingredients required by the animal frame, and hence possesses the sanction of science as well as of custom.

311. Different kinds of diet are required by *different habits* of life.

Under some circumstances, men need a more than usually *nutritious food:* under others, they need a more than usually *heating diet.*

312. Great exertion calls for *very nutritious* food.

Since all activity is attended by waste of structure (258), it is clear that the more exertion there is made, the more nourishing material must be furnished to keep the organs in a working state; but this nourishing material must be of a plastic, rather than of a heating kind. If large quantities of oily matter were taken, under these circumstances, the system would become either oppressed with combustible matter, or with the heat resulting from its burning, instead of being strengthened for the increased labor. The best addition that can be made to the ordinary diet, to meet this requirement, is lean meat. One pound of meat contains as much plastic matter as two pounds of bread or four pounds of potatoes. Lean meat is the most nutritious, and the least heating of all the substances employed as food. A plentiful meat diet has the same effect upon a hard-working man, that a corn diet has upon a hard-working horse; it fits him to bear the constant strain made upon his muscular structures, and it does this without overloading his body with combustible matter at the same time.

313. Absolute repose necessitates the employment of a *chiefly farinaceous* and a *very sparing* diet.

When very little exertion, of either body or mind, is made, the waste of the organized structures is proportionally slow; and, hence, very little plastic food is called for. The farinaceous materials of the nature of rice, potatoes, and tapioca, then, contain as much direct nourishment as the inert system requires for its sustenance; and they have the further recommendation that they are, at the same time, not of a very heating nature, because their starch needs to be converted into oil in the system, before it can be burned, or produce heat.

314. Exposure to great degrees of cold is best borne when a *very oleaginous diet* is used.

The Esquimaux, who habitually brave the extreme cold of the Arctic winter, live almost exclusively upon seal oil. Two ounces of oil produce as much effect in heating the body, when consumed as food, as an entire pound of lean meat.

315. Farinaceous foods are most suitable for *hot climates and seasons*.

In the hottest regions of the earth, the mean temperature of the air is not sufficiently high for the purposes of the animal body; hence, the internal furnace is kept burning in them, as well as in the colder climes. But less fuel, of course, needs to be consumed. Under such circumstances, starch forms a better fuel than oil, because it is less combustible and burns more slowly. Oil differs from starch, principally in containing ten times less oxygen; it therefore has a greater attraction or thirst for the corrosive element, in this degree, and of necessity gives out more heat, when burned. Two ounces of oil produce as much heat in the body as five ounces of starch; but the heat produced by the burning of the starch is also set free more gradually, because this principle has to undergo a preliminary process of conversion into oil, before it is ready for use. It is on this account, that the natives of India and China find rice so suitable an article to form the chief bulk of their food.

316. In temperate climates, the diet is best *varied with the seasons*.

The remarks made in relation to the various heating and nourishing powers of different kinds of food, apply as much with regard to seasons, as they do with regard to climates. More oleaginous principles are required in winter than in summer. It frequently happens that persons who take cod-liver oil through the winter, in England, get so oppressed by it in summer, that they are constrained then to refrain from its use.

217. Animal food is not *absolutely essential* for the support of life.

This may be inferred from the fact that there are animals which feed exclusively upon vegetable matters. The ox, that supplies beef to man, eats only grass and turnips. Many individuals of the human species take nothing but vegetables as food, and yet preserve vigorous health for long periods of time. There is no nutritious principle in meat that is not also found, although in a more sparing and less condensed form, in vegetable substances.

318. A *mixed animal and vegetable diet* is the best adapted to the general wants of man.

It is a debated question whether man is designed by Nature to feed on vegetable or animal substances. It is not difficult, however, to find a satisfactory answer to this. He is intended by Nature to live on both. He is *omnivorous* (*"devouring all things"*) in the widest sense

F*

of the term. He eats, and thrives upon all kinds of food, and may be restricted to an exclusively animal or vegetable diet with impunity, provided only a due proportion of plastic and combustible principles are supplied, in the condition in which they are available for use. In tropical lands, man luxuriates in delicious, sugary fruits, and in other productions of the ground; in temperate climates, he mingles bread and meat in various proportions; on the wide prairies, he can get nothing but buffalo beef and venison; and in the dreary, arctic waste, fish and seal oil are his sole resources;— yet, under all these varieties of circumstances, he still manages to keep his frame in healthy vigor, and fit for its work. Man, in reality, is enabled to exist upon a great diversity of food, in order that he may dwell in a great diversity of conditions; his omnivorous capacities have been conferred upon him, in order that he may "subdue the earth" and cover its surface, from the luxuriant tropics to the desolate poles, with his race. In civilized and densely-peopled lands, where mouths multiply much faster than the natural productiveness of the soil, a mixed diet of animal and vegetable substances is invariably adopted, because, in practice, it is found to be both more economical and convenient. It proves to be easier to make the land yield an augmented produce, under the application of science to its culture, when that produce is taken out in mutton, beef, and grain, than when it is procured in the form of grain alone. Every possible advantage has then to be sought out and seized upon, in order that the rapidly-increasing numbers of the people may be comfortably and sufficiently fed. But the employment of the mixed diet, by highly civilized races of mankind, has another very important advantage. It enables a larger quantity of plastic substance to be thrown into the system at any time, to answer a special purpose, without increasing, in the same proportion, the quantity of fuel present in the frame. It furnishes a means for strengthening the body of the hard-working man for extraordinary exertion, without its being heated in a corresponding degree. The natural food of the horse is grass, but so soon as the horse is taken from the pasture and set to drag heavy loads, it is found that its muscular powers must be sustained by the addition of a certain quantity of corn to its daily fare; for the corn has much more plastic nourishment in it than an equal quantity of grass. Meat is added, for the same reason, to the toiling man's daily fare; that is, in order that the excessive waste of plastic substance, entailed by his labor, may be promptly and easily repaired. These remarks apply as much in the case of mental as of bodily exertion. The best physiologists are of opinion, that although there can be no doubt an exclusively vegetable diet is sufficient for the production of a full and perfect development of the bodily frame of man, it is equally clear that the addition to it of some animal food favors the formation of the highest power of mind.

319. The food of man becomes more easily digestible by being *cooked*.

Man may be distinguished as the *cooking* animal. The savage, in his rudest state, makes a fire, and prepares his food, by roasting or baking it, before he eats it. No other species of animal performs this operation. All the diversified processes of the cook's art have one object in view (so long as they are confined to their lawful province), the reduction of the various nourishing principles of the food into their most soluble condition. Cooking, indeed, imitates many of the actions that are naturally brought about in the digestive canal, and may be viewed as a preliminary stage of digestion. It breaks the films of the starch grains, and converts them into a sort of sugar or gum; it softens and opens out the texture of the glutinous and albuminous matters: and it changes the oily principles into bland, milk-like emulsions. It is best that no uncooked food, excepting ripe fruits, should be eaten, and they can hardly be viewed in the light of an exception, because they have really undergone a sort of cooking in the heat of the sun.

320. Savory admixtures of food are addressed more to *tempting the appetite*, than to rendering the food easy of digestion.

On this account, the refinements which modern luxury has introduced into the cooking art, are evils rather than benefits. The mingling of savory foods into varied dishes that gratify the palate, replaces natural and healthy appetite by unnatural and disordered craving. All the appetites are given to animals to ensure the fulfilment of certain actions that are essential to the well-being of the body. Eating, in common with many of the other ordinary operations of life, has been made pleasurable, in order that the creature may be induced to take the food that is necessary for its sustenance, for the sake of the enjoyment that is attendant upon the act. When, however, numerous different kinds of food are offered to the palate in succession, and when highly seasoned and richly-flavored dishes are presented in the place of simple fare, men are apt to continue eating for the prolonged gratification of the sense of taste, long after the real wants of the system have been satisfied; and so the stomach is oppressed with a load that is far beyond its powers of management, instead of being only fairly tasked with the work that it is easily able to accomplish. Persons who value the blessing of uninterrupted health, should always train themselves to make their meals consist of at most one or two dishes of simple and plainly-cooked food. All indulgence in the pleasures of the table, beyond this, is playing with temptation, and planting seeds that are almost sure to ripen into future suffering and evil.

321. *Hunger* is the natural appetite for food.

Hunger is an uneasy sensation, which seems to arise in the stomach after a certain amount of abstinence, but which really originates elsewhere, for it is capable of being appeased by the introduction of nutritious matter into the system, by other means than through the stomach. It is the sense of emptiness in the frame at large. Nature indicates the fact, in this way, when the plastic material of the body is near to exhaustion, and needs to be recruited. Hence, so long as the appetite is not pampered by highly seasoned and varied foods, hunger may be taken as the safe guide as to the times when, and extent to which, refreshment is necessary: so soon as the sense of hunger is appeased, no more food should, on any account, be swallowed. But in order that this friendly monitor may have due attention paid to its promptings, it is essential that the meals should be taken slowly and with deliberation. The process of digestion requires some time for its performance. Nourishment is not introduced into the system as rapidly as food may be into the digestive organs. Hence, hunger may continue to be felt, even after the stomach has been loaded to repletion, because the system may be yet empty of nourishment, although the stomach is full of food. When the meals are taken slowly, enough nourishment for the blunting of the sensation of hunger gets into the system by the time the stomach has received a sufficient charge of food.

322. Successive meals should be separated from each other by *intervals of about six hours.*

It takes about four hours to complete the digestion of an ordinary meal; but the stomach should never be called upon to work again, immediately upon the completion of its task. It needs, at least, a couple of hours for repose. It is a very good rule to make three meals in the day, with intervals of about six hours between them, and with a longer period of twelve hours' rest at night. This affords the healthy stomach the opportunity of digesting as much food as will prove amply sufficient for the renewal of any waste the body can experience.

323. Human adults require about *two pounds* of dry food daily, to repair the waste of their frames.

This presumes that the food is of a tolerably nutritious kind ; such for instance as is afforded by a judicious admixture of bread, meat, potatoes, and butter. Two pounds are then sufficient for men who are supporting a considerable amount of exertion. It has been found that 26 ounces of dry vegetable substance, and 10 ounces of meat (36 ounces in all) constitute an ample daily allowance for the preservation of the health of sailors, exposed to the vicissitudes and labors of service, on board ships of war. In the experience of the Edinburgh House of Refuge, 23 ounces of solid food (including one ounce of meat and seven ounces of bread) proved hardly to be a sufficient diet for long continuance. A life of idleness may, however, be sustained upon this stinted allowance very well. Twenty-five ounces and a half of solid food are ascertained to constitute a sufficient daily dietary for people who are confined in inactivity in the Union-houses. There is one instance on record, in which a man lived for 58 years upon only 12 ounces of solid food, principally of a vegetable kind, taken daily, and yet continued in perfect health. A fair diet for all the ordinary purposes of life is afforded by half a pound of wheaten bread, taken with a little butter at breakfast ; half a pound of well-cooked meat, with a quarter of a pound of bread, and half a pound of potatoes, or other dry vegetable substance, at dinner ; and half a pound of bread, with butter, for the evening meal. The *exact quantity of food* that can be taken with benefit, of course depends very much on age and the habits of life ; but most men of mature growth would find it to their advantage to diminish rather than to increase the allowance specified above.

324. The stomach can only digest a *fixed quantity* of food in a given time.

The amount of gastric juice and saliva that is furnished, is determined by the wants of the body, and not by the quantity of food that is placed in the stomach. Somewhere about a pint of saliva and three pints of gastric juice are formed every 24 hours, when the digestive functions are in vigorous action, but these four pints of solvent liquid can only dissolve a certain amount of food (a little more or less, according to its nature). If, then, more food than this is habitually placed in the stomach, portions of it must remain undissolved, and the undissolved portions must either be got rid of by some unusual action of ejection, or they must remain oppressing and irritating the offended organ.

325. Sickness is the natural means whereby an overloaded stomach *gets rid* of food which it cannot digest.

In sickness, the ordinary movements of the stomach and bowel, by which the digesting food is carried onwards, are reversed. So that the contents of the stomach are thrown into the mouth, and the contents of the upper part of the bowel are thrown into the stomach. Whenever sickness is long continued, bile is almost sure to make its appearance amidst the rejected matters, but this is the necessary result, and not the cause, of the vomiting.

326. Indigestion is the discomfort caused by *the presence of indigestible food* in the stomach and bowel.

When the stomach is habitually loaded beyond the extent of its powers, its action gets deranged, and becomes weak and painful. All feeling of appetite disappears. Uneasiness and a sense of sickness ensue, whenever fresh food is placed in the organ. Heartburn, cramp, griping, headache, sleeplessness, and palpitation of the heart follow, provided the overtasked organ is not allowed time to right itself. When the stomach is weakened by unfair usage, food will sometimes remain undissolved in it for entire days. In the great majority of cases, the suffering of indigestion is caused by the stomach having been asked to do work that is beyond its capacities and strength.

327. *Abstinence is the natural cure* for indigestion.

There is but one cure for indigestion. That which alone could have prevented the mischief is also the only influence that can effect its removal, when once it has been produced. Moderation in eating and drinking, pushed almost to abstinence, will reclaim a disordered stomach to its healthy condition, in ninety-nine cases out of a hundred, provided the habits of injurious indulgence have not been sufficiently long persevered in to have produced derangement of structure in any important part of the apparatus. This moderation is all Nature wants, when she begins to utter her complaints, in the form of symptoms of disorder ; but this is the only concession she can accept.

328. Habits of *self-denial and self-control* in eating are essential to the preservation of health.

One of the best lessons that can be learned in early life is the wisdom of acquiring powers of self-denial and self-control, in all matters where the gratification of the appetite is concerned. Every act of self-indulgence, that is committed in opposition to the dictates of prudence, is a transient pleasure purchased at the expense of future pain. Every resolution of self-denial that is carried into effect, in defiance of temptation to sensual enjoyment, is another step gained in advancing towards happiness. These truths are very forcibly illustrated by the subject under immediate consideration. Every one who desires to possess uninterrupted health will do well to train himself to live on the plainest and simplest food, taken slowly and in moderate quantities, with intervals of at least six hours intervening between the successive meals.

329. The human frame can endure *complete abstinence*, in matters of food, for very lengthened periods of time.

Man's body seems to be more tolerant of deficiency

in diet, than it is of excess. Complete abstinence from every form of food and drink can be supported for eight or ten days before death ensues, and the fatal result is put off, for even a longer period, if only a little water is swallowed occasionally. During this time, however, the body of course wastes from day to day. The structures go on consuming themselves, although their waste is no longer repaired.

CHAPTER XIII.

THE BLOOD.

330. The dissolved food is *collected into one receptacle*, after it has been absorbed from the bowel.

This is called the "receptacle of the chyle." Each little vessel, that originates in an absorbing villus of the bowel (296), joins others coming from other villi. In this way, enlarging trunks are formed which run along upon the coats of the bowel, and are termed "lacteals," because they carry a liquid that looks like milk (*lac* is the Latin word for milk). The lacteal vessels, that issue from different parts of the bowel, all get gathered together into a web of membrane, which serves to hold the several folds and doublings of the canal in their proper positions. This connecting web, that thus envelopes and sustains the twisted digestive tube, is spread out from the lower part of the back-bone, something after the fashion of a crumbled-up fan. The back-bone serves it as a fixed point of support, and its outer expanded edge is attached to the twisting folds of the bowel ;—it is called the "*mesentery*," (a word derived from two

Greek terms, *mese*, " *the middle*," *and araia*, " *stomach.*")
The lacteal vessels run from the bowel along this fan-
shaped web (taking the place of the rays of the fan),
until they all meet together near its attachment to the
back-bone, and so collectively form there the receptacle
into which the chyle is poured.

331. **The dissolved food is conveyed along
a tube laid down for the purpose, until it is
poured into one of the large blood-vessels,
and so** *mingled with the blood.*

The vessel which carries the chyle up from the chyle
receptacle into the blood, is called the thoracic duct, because it runs through the " thorax," or chest. (" *Thorax*" is the Latin word for chest.)　It begins at the receptacle of the chyle, passes along just in front of the back-
bone, quite up to the neck, and there ends in one of the
large veins.　Where it ends in the veins, there is a little
trap-door, or valve, which allows the chyle to flow freely
down into the blood-vessels, but prevents all return of
the blood into the duct that carries the chyle.

332. *The blood* is formed *out of the chyle.*

If it were not for the fresh supplies of nutritious matter that are poured into the blood from time to time,
this liquid would soon be exhausted by the incessant calls
that are made upon it from the vital organs at large.
Whenever the flow of chyle to the blood is interrupted,
the body soon becomes bloodless and exhausted.　The
wasting disorder, familiarly known as " Mesenteric disease," is due to an obstruction of the chyle ducts, occurring in the web of the mesentery.　It is a kind of starvation caused, not by the want of food, but by the inability
of the dissolved food to get from the digestive organ into
the blood.　The word " *blood*" is derived from the Anglo-
Saxon, " *blœdan*," to bleed.

333. Blood is *chyle in an altered state.*

Since blood is formed out of the chyle, it, too, is composed of the nutritious principles of the food (295).　In

the blood, however, these principles are not only liquefied, for the convenience of easy transport, and introduction into the small cells and spaces of the animal body, but they are also advanced a stage further towards the final state of organized structure. Blood is food just in the act of being made alive. Although blood is composed of the same material as chyle, it nevertheless differs from it in several important particulars.

334. Blood differs from chyle in being of a deep red color.

The chyle is a milk-white liquid, but the blood, which is made out of chyle, is of a deep purple, or scarlet color (*excepting in cold-blooded animals*). This color appears gradually, as chyle is perfected into blood. The chyle has a faint pinkish tinge at the top of the thoracic duct (331), before it is poured into the vein in which this vessel ends.

335. The color of the blood is caused by the presence of an immense number of *small, red particles*, that float in the liquid.

Fresh blood may be so poured upon porous paper that it runs through as a clear transparent liquid, all the red portion remaining behind as a deeply-colored mass. When a drop of blood is placed between two pieces of glass, and examined by a magnifying lens, a host of small red specks is seen floating in clear liquid. The blood looks, to the unaided eye, like a uniform, red liquid, because these specks are not transparent, and because they are closely crowded upon each other in countless myriads.

336. The small, red particles of the blood are called *blood corpuscles*.

In other words, the "*little bodies*" of the blood. ("*Corpusculum*" is the Latin term for "*little body*.")

337. The clear fluid, in which the colored blood corpuscles float, is called the *blood plasma*.

The word "*plasma*" is derived from the Greek term (*plasma*), which means "work" or "workmanship." The plasma of the blood is the perfected, plastic material that is employed in all the constructive work of the system. This plasma is also called the "blood liquor," (*liquor sanguinis*).

338. The blood corpuscles are really minute *living cells*.

When a drop of blood is placed between two pieces of very thin glass, and examined by a powerful microscope, the corpuscles are seen to be hollow bladders of filmy membrane, containing liquid in their interiors. The figure (*Fig.* 27) represents the appearance of the corpuscles of the blood, when magnified some half-million of times.

Fig. 27.

339. the blood corpuscles are *exceedingly minute*.

This will be at once admitted, when it is understood that about half a million of them might be laid upon each of the figures sketched above. It would take fifty thousand of them to cover the head of a small pin. There are no less than three millions of them in such a little drop as may be held suspended on the point of a needle! It would be quite impossible to express in numbers the myriads of millions of them there must be in the entire blood of a full-grown man.

340. The colored blood corpuscles are of a *flattened figure*.

They are four times broader than they are thick. In figure 27, some of them are seen edgeways at *a*. They have a little pit or dimple impressed upon each of their sides. Their shape is thus something like that of the body which would be formed, if two flattish watch-glasses were placed together by their rims, and were then so held between the finger and thumb, and squeezed in by sudden pressure where they were held. The appearance presented by these pits is shown in the circular outlines of the figure.

341. The films of the blood corpuscles are *devoid of color and nearly transparent.*
Hence their contents can be seen shining through these outer transparent walls. •

342. The blood corpuscles contain an *opaque, red liquid,* upon which the name of *hæmatin* has been conferred.
The red color of the blood is really due to the liquid substance contained within its corpuscles. The word " *hæmatin*" is derived from the Greek " *aima,*" which signifies " blood."

343. Hæmatin is a highly complex substance, composed of the *four organic elements* (122).
Hæmatin is a very complex body, something of the nature of gluten (239). It is, indeed, the glutinous principles of the food, exalted towards animalization. Each molecule of hæmatin seems to contain within itself forty-four atoms of carbon—twenty-two atoms of hydrogen—three atoms of nitrogen, and six atoms of oxygen. Its composition may, therefore, be represented symbolically by the annexed diagram. It appears to be glutinous substance, from each of whose molecules small portions of carbon and nitrogen, and large portions of oxygen and hydrogen have been removed. There is one other peculiarity, however, about hæmatin. It is always closely associated with a very considerable proportion of iron, which can, nevertheless, be withdrawn from it by chemical operations, without the destruction of its existence. The red liquid of the blood corpuscle contains as much as six per cent. of iron.*

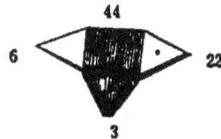

* There are really two different kinds of complex bodies mixed together in the liquid of the blood corpuscles ; one called *hæmatin,*

344. Hæmatin is made out of the blood plasma (337), *by the agency of the living blood corpuscles.*

It will be remembered that one of the operations of cell life is changing the nature of the liquids that are passed through the films of its vesicles (169). When certain portions of the plasma, in which the blood corpuscles float, are imbibed through their films by osmose (163), those portions undergo a re-arrangement, so far as the constituent atoms are concerned, and become the rich, red liquid which is known as hæmatin. As a general rule, living cells make the matters they contain, and it is a chief part, indeed, of their duty to carry on this manufacture. The great office, which the innumerable corpuscles that float in the blood have to perform, is the preparation of the red liquid upon which its color depends.

345. The red hæmatin, prepared by the blood corpuscles, is the especial *food of the important structures* **that enable animals to move and feel.**

Hæmatin is the perfected food of *animal life.* It is the substance which is organized into muscles and nerves. The general plasma of the blood is suited for the formation of most of the textures of the body, but it is not equal to the construction of those textures which are intrusted with the accomplishment of the final results of animal organization, motion at will, and perception of external objects and relations. Consequently, a myriad of little workmen are provided to fit the plas-

---

and the other called *globulin.* They are, however, of somewhat similar nature, and, therefore, for the sake of simplicity, may be spoken of as one body, under the general denomination of *hæmatin.* In strict accuracy, the two are distinct, and can be separated by artificial means. They are then found to be different bodies, because the hæmatin is red, and the globulin devoid of color. Hæmatin and globulin, when mingled together, as they are in their natural state during life, are designated " *cruor.*" " *Cruor*" is the Latin name for "blood," such as is poured out from a wound.

ma for this higher office : these workmen are the colored
blood corpuscles, and the red hæmatin is the production
of their activity. Whenever there is great abundance
of red corpuscles in the blood, the offices of animal life
are vigorously performed; but whenever there is any
deficiency of them, all the muscular and nervous operations are languid and imperfectly carried on.

346. Some of the blood corpuscles are *pale
and devoid of color.*

These colorless globules are seen floating in the blood
plasma amidst the colored ones.
They are nearly spherical, instead
of being flattened, and are not in-
dented at their sides, but are
dotted with little grains instead.
The figure (*Fig.* 28) represents
at *a* the appearance of one of the
pale corpuscles, magnified about
a million and a half of times.

Fig. 28.

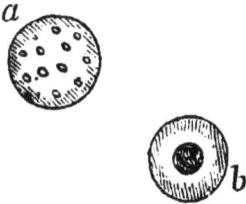

347. The pale corpuscles of the blood are,
in all probability, *red ones in the process of
formation.*

The pale corpuscles are far less numerous in the blood
than the red ones; there are generally not more than
one of the former to every fifty of the latter. Comparatively few of them are seen, because they are changed
into colored corpuscles soon after they are produced.
They may be watched with the microscope, passing
through various stages of transition, until they get
finally changed into complete red corpuscles. During
these stages of progression, they get constantly smaller,
firmer, heavier, and more colored, until at last they are
flattened and pitted on their sides. The sketch at *b*
(*Fig.* 28) shows the appearance of the corpuscle when
the change is complete; the representation being magnified a million and a half of times.

348. The pale corpuscles are *formed out of chyle.*

The pale corpuscles are formed in the chyle vessels. They do not, however, appear in the chyle either of the bowel, or of the chyle ducts near to it. They only present themselves in the larger vessels of the mesentery (330), and in the thoracic duct (331).

349. The pale corpuscles are formed *by the agency of little organs called glands.*

The lacteal vessels are enlarged into little knots in several places, during their progress along the mesentery. These knots are called the "*mesenteric glands.*" (The word gland is derived from "*Glans*"—the Latin for *an acorn or nut,*—and was conferred as a name, upon these organs from a fancied resemblance to the thing so termed). The tube of the lacteal vessel branches into several divisions, where it passes through one of these glands, as represented in *Figure* 29, and these are then lined by a thick layer of loose cells, which are identical in nature with the pale corpuscles of the blood. These clustering cells multiply by one of the ordinary processes of cell division (173–176), being fed by the chyle which is brought to them by the lacteal vessel, and the several generations, produced by the multiplication, are cast loose into the lacteal vessel, and sent onward as pale corpuscles towards the thoracic duct, and the blood. No pale corpuscles are ever found in the lacteal vessels before they have passed through the mesenteric glands, but they make their appearance immediately beyond.

Fig. 29.

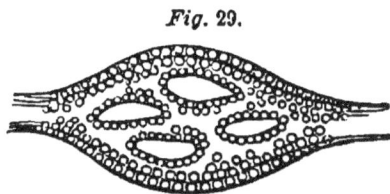

350. The colored corpuscles of the blood *die as soon as their work is completed* (264).

As soon as each little cell has formed its contribution of hæmatin, and has conveyed it to some muscular or

nervous structure, requiring renovation, it bursts, or is
dissolved in the blood plasma, and so disappears for ever
There is thus a constant production and destruction of
blood corpuscles going on within the frame. Innumer-
rable as they are, fresh myriads of them are being called
into existence at every instant, and old myriads of them
are vanishing from the scene. In the earliest stage of
individual life, before the young creature has begun to
take food for its support, the red corpuscles multiply
by subdivision, but subseqently they never do so. They
are then exclusively formed as pale corpuscles, out of
the chyle in the mesenteric glands, develope into red
corpuscles, form their hæmatin, turn over what they form
to muscle or nerve, and then end their lives in decay.
The blood corpuscles afford an illustration of cells that
change their characters subsequently to their first production, in order that they may become fit for the performance of a particular office (178). When this takes
place, the cell usually loses the power of multiplying itself by the production of offspring. All its vital power
is expended in the accomplishment of its transforming
task.

351. About a *seventh part of the blood* is composed of corpuscles.

In seven pounds and three-quarters of blood, there is
generally about one pound of corpuscles. But this proportion varies to a considerable extent in different individuals. Excessive exertion of body and mind quickens
the destruction of the corpuscles of the blood, and therefore tends to diminish the proportional quantity of them
present in the liquid. An abundant and nutritious diet
quickens their formation, and tends to increase their
number.

352. The plasma of the blood (337) is liquid, so long as it is retained *in the living vessels*.

The blood corpuscles are rolled along in streams of
the liquid plasma, to all the several parts of the frame

that stand in need of their sustaining influence. The greater part of the blood is still kept in the liquid condition for the convenience of transport. It is food in the final stage of preparation for organization, but as yet fluid, in order that it may be poured to every part of the frame where the work of organization is going on.

353. The plasma of the blood *separates into two portions*, one solid and one liquid, so soon as it is removed from the living vessels.

When blood is drawn from a vein and allowed to stand in an open vessel, for a few minutes, a firm mass collects together in it, and floats up to the surface of a thin, clear fluid.

354. The solid, which forms in blood plasma, when it is drawn from a vein, is called *the clot*.

" *Clot*" is derived from the old German word " *kloss*," which means a "clod." The Latin word for the same thing, also in frequent use, is " *coagulum*" (243).

355. The part of the blood plasma which remains liquid, after the clot has curdled and separated from it, is called *the serum*.

" *Serum*" is the Latin name for " whey." The serum of the blood plasma is a thin transparent fluid, very much resembling the whey of milk.

356. When the clot forms in blood plasma, after it has been drawn from a vein, it *entangles all the corpuscles* in its mass.

Hence, the clot of the blood plasma is of a deep red color, and opaque, and the liquid that is left behind is clear and devoid of color.

357. The clot of the blood plasma is *colorless in itself*.

All the blood corpuscles may be washed away from the clot of the plasma, by frequent rinsings with water. There then remains behind only a whitish, or ash-colored mass.

358. The clot of the blood plasma possesses *a fibrous texture.*

When the clot is examined by a microscope, it is seen to be composed of a net-work of interweaving fibres. It is in the meshes of these fibres that the corpuscles are caught and entangled, when the coagulation first takes place.

359. The clot is principally composed of a peculiar, dense substance, which is called *fibrin.*

The word *fibrin* is derived from "*fibra,*" the Latin term for a thread. It is so called on account of the thready way in which it is deposited when it coagulates (357).

360. Fibrin is a *highly plastic* substance, of a very complex nature, and composed of all the four essential, organic elements (122).

Fibrin is so nearly of the same composition as gluten (239–240) that chemists cannot find any difference between the two bodies. Its molecule may therefore be symbolically represented in the same way as the molecule of gluten (239). Fibrin is really *animalized gluten.* It is the complex product, which the plant has formed, rendered a trifle more adhesive and plastic, to adapt it for purposes to which it is about to be applied in the animal economy.

361. Fibrin makes the blood *adhesive and viscid.*

Blood is a thick, gummy liquid, and not a thin one, like water. A layer of it leaves a coat, like varnish, when it dries. The thickness and viscidity of blood are due to the fibrin that is present in it. Wounds heal be-

cause their sides are glued together by means of the adhesive fibrin, which is left upon them during the continuance
of bleeding.   Whenever fibrin is deficient in the blood,
in consequence of disease, wounds can hardly be got to
heal.   The thickness of the blood also makes it run more
freely.   It has been ascertained that liquids, possessed
of the thickness of the blood, flow through fine vessels
and tubes more easily than water itself.   When there is
less of fibrin in the blood than the proper proportion,
this liquid has a tendency to escape through the walls of
the vessels, and so to cause bleedings that can hardly be
checked.

362. A very *small quantity of fibrin* is sufficient to make the whole mass of the blood
adhesive and viscid.

In ordinary health, there is not more than half an ounce
of fibrin in sixteen pounds of blood.   This small quantity, however, is amply enough to make the whole liquid
adhesive, when it is diffused through it.   The comparatively large clot that forms in blood, when drawn from
a vein, has its *bulk* due to the corpuscles, and portions
of the other constituents of the plasma that are entangled
in the adhesive fibrin.   The fibrin is the coagulating principle of the mass.   In inflammatory disorders, the blood
has two or three times more than its usual quantity of
fibrin.   It then becomes too thick and tenacious to flow
freely, and stagnation is consequently very apt to result.

363. Fibrin seems to be formed by the
agency of the colorless corpuscles.

The colorless corpuscles appear to form tenacious
fibrin out of the blood plasma, as the colored ones form
hæmatin.   The few colorless corpuscles left unchanged
amidst the colored ones (346) probably are engaged in
carrying on this work.   As the colorless corpuscles are
imperfect colored ones, the fibrin may also be looked
upon as a sort of imperfect hæmatin.   Since so small a
proportion of fibrin is sufficient to render the entire mass
of the blood duly tenacious and viscid, there is an obvi-

ous reason why the colorless corpuscles are so much fewer than their red companions. The fact that fibrin is formed by the colorless corpuscles is shown by its increasing in quantity in the blood, whenever the colorless corpuscles augment in number, and by its diminishing in quantity when they decrease. The chyle possesses no coagulating power before it reaches the lacteal glands, but it acquires the power after it has passed through them, at the same time that it receives pale corpuscles from them (348).

364. The coagulation of the fibrin of the blood is a *sort of rude organization.*

The clot of the blood plasma is formed of fibres which are woven together in a definite manner. This arrangement is evidently the work of organizing power, for precisely the same is found in certain structures that are made out of fibrin within the body.

365. The fibrin of the blood plasma is destined for the formation of certain structures that are needed for *mechanical purposes,* and that therefore *possess but small vital and formative powers* within themselves.

It is necessary that the material out of which these structures are formed should be very plastic and tenacious, as it has to make up by its own proneness to get formed into structure, for their low capabilities for selecting and attaching to themselves new substance. The gristle and tendons by which bones are connected together, are illustrations of structures of low vitality that are formed out of fibrin. The first step in the repair of injuries that have been produced by accidental violence, is the deposit of a layer or mass of tenacious fibrin. This principle is the great plastic agent of the economy. It renders the blood duly thick and adhesive. It is moulded into all sorts of fibrous apparatus, that would be unfit for mechanical service if they were composed, like vital organs, of living cells. And it is the great rcsourcc to which nature looks for reparative aid in all cases of urgent need.

366. Blood is composed of three different parts—*corpuscles, fibrin, and serum.*

In living blood, the fibrin and serum arc mingled together as liquid plasma in which the corpuscles float. In dead blood, the corpuscles and fibrin are mingled together as clot, which scparates from the serum. All this may be expressed to the eye at a glance, in thé following way :

Living Blood consists of	*Dead Blood* consists of
Corpuscles............... Corpuscles ⎱ form Clot and ⎰ Fibrin ... ⎰ and Plasma—consists of...... ⎰ Serum.........Serum	

367. A white deposit is formed in the serum of blood when it is heated to the *temperature of boiling water*.

The fibrin of the blood separates from the plasma, when it is withdrawn from the body, as a solid clot, at the ordinary temperature of the air. But there then remains in the liquid scrum another solid substance, which also separates as a clot, so soon as it is heated to 212 degrees of Fahrenheit's scale.

368. The white clot, which is formed in blood serum at a boiling temperature, *is the substance already described as albumen* (242).

The albumen found in the blood is perfectly identical with the albumen formed in vegetable juices. It is a complex substance made of the four essential elements (243). It differs from gluten and fibrin only in its being soluble in water at the temperature of the living body. It is therefore less plastic than they are. It manifests no tendency to become consolidated into structure. White of egg is albumen mingled with water.

369.. The albumen of the blood is the *plastic principle* of the food in the simple

and crude state in which it is introduced into the body after digestion.

The albumen of the blood is not made in the animal body, as the hæmatin and fibrin are. It is merely the plastic and organizable substance formed by the vegetable, dissolved and introduced into the economy of the animal. The hæmatin and fibrin are made in the blood, out of and at the expense of the albumen. The corpuscles and fibrin of the blood possess a sort of life of their own, but the albumen has no vitality. It is merely the reserve store of nutrition upon which nature draws, when she prepares the more advanced materials of organization.

370. The albumen of the blood is *twenty times* more abundant than the fibrin.

There are about ten ounces of albumen in sixteen pounds of blood. It is necessary that there should be much more of it than of fibrin, because all the structures of the body are primarily formed out of it, while only a few textures of low vitality are made out of the fibrin.

It will now be understood that the albumen of the blood, which remains dissolved away in the serum, is the crude condition of the nitrogenized principles of the food, ultimately designed for plastic and constructive purposes; while the fibrin and the hæmatin are the same principles perfected for the purpose to which they are to be applied, through the influence of the living corpuscles. Albumen is the raw material out of which fibrin and hæmatin are manufactured. Fibrin and hæmatin are albumen vitalized and just about to become living structure.

371. A small quantity of *oil* is always contained in the serum of the blood.

The blood contains about the same quantity of oil that it does of fibrin, namely, half an ounce to every sixteen pounds of the liquid. This principle exists in the condition of very minute globules, encased in thin, filmy coverings of coagulated albumen. In this way it

is prevented from running together into drops. It is
principally introduced ready formed from the food.
The whiteness of the chyle is due to the presence of
numbers of these albumen-covered oil globules, which
are suspended in its watery parts. Oil is, however,
sometimes formed out of the albumen of the serum.

372. The oil of the blood is principally
used *as fuel* in keeping up the warmth of the
body (303).

The oil of the blood is burned in the body, by means
of an arrangement that will be described hereafter
(chap. xxiii.). Through this burning, heat is set free
and used in warming the organs to the temperature essential to the performance of their functions. Still
smaller quantities of sugar are also found associated
with the oil. These are consumed in the same way, and
to the same end.

373. The oil of the blood also takes some
part in *forming the organized textures.*

The presence of oil seems to be essential to the first
stage of the formation of cells (236). The oil granules
of the chyle (371) assist, no doubt, in the formation of
the colorless corpuscles (349). The greater part of the
oil of the blood, that is destined for constructive work,
is, however, mixed up with the hæmatin of the corpuscles.
The purpose to which this is put will be hereafter seen.

374. Superfluous oil is removed from the
blood, *and stored away as fat.*

A certain amount of combustible oil must be present
in the blood (371). This is mostly procured directly
from the oleaginous principles of food. But if, at any
time, this source of supply runs short, some of the albumen of the serum is transformed into oil. If, on the
other hand, the food supplies more oil than is needed for
present use, the superfluous portion is taken out of the
blood and stored up in little membranous bags, whence
it may be again drawn, if any future need should arise.

Fat consists entirely of little membranous bags of oil, and is of no *immediate* use in the economy.

375. The serum of the blood contains all the *inorganic principles*, that are abstracted from the food to be employed in constructive service (126 and *seq.*).

These inorganic principles, it will be remembered, are mainly sulphates of potash and soda—phosphates of potash, soda, magnesia, and lime—chloride of sodium (*common salt*) and iron. They are all in the state of saline compounds (152), and are held in complete solution in the serum. There are about two ounces of such inorganic principles in every sixteen pounds of blood. If a small quantity of blood is dried and burned, all its organic constituents (hæmatin, fibrin, albumen, oil, and sugar) are decomposed and driven off, and the inorganic constituents alone are left behind as an incombustible ash.

376. The inorganic constituents of the blood *are used in the construction* of different textures of the frame.

They are, however, subordinate and special, rather than general and essential, ingredients of the textures (127). Some go to form one kind of structure or substance, and others help to constitute structures or subtances of a different sort. Thus the phosphate and carbonate of soda help to keep the albumen soluble in the serum, and assist in forming the bile. Potash is furnished to the muscular substance. Iron enters into the composition of the muscles, the red blood corpuscles, and hair. Lime consolidates the bones and the teeth; and salt (chloride of sodium) exists in many of both the solids and fluids of the organic frame.

377. More than *three-quarters* of the blood *is pure water*.

If all the water be driven off from sixteen pounds of

blood, by gentle evaporation, there will remain about three pounds and a quarter of solid residue. This consists of all the cell films of the corpuscles—hæmatin, fibrin, albumen, and inorganic ingredients. In sixteen pounds of blood, no less than twelve pounds and three-quarters are water alone.

378. The *free mobility* of the blood depends upon the large proportion of water that enters into its composition.

The blood is the several principles of nourishment designed for the support of the various living organs, mingled together and kept in a liquid state, for convenience of transport to the places in which they are needed. This liquidity is effected by the presence of water. Water thus performs the same office for animals that it does for plants. It carries for them the various solid substances that they need as food (91).

379. Water also favors the *accomplishment of the various changes*, that are connected with the production of vital actions in the several organs of the frame.

The various changes of substance that are intimately connected with the transformation of organizable principles into organized material, and with the production of power out of decomposition (258), are rendered more easy of accomplishment when the different substantial agents that are to influence each other, are kept diffused or dissolved in water, because then the molecules of the two can be readily brought into very close and intimate communication with each other. The presence of the due proportion of water in the blood also sets up the action of osmose, by which the carrying power of the liquid is called into play. The quantity of water that the blood contains, varies, within a small limit, according to the temperature of the air, the exercise taken, and the nature and quantity of the food and drink; but it is nev r allowed to depart very far from the natural standard

(79 parts of water to every 21 of solid ingredient). The contrivance that is adopted, for preserving this due proportion of water in the blood, under all the varying circumstances and habits of human life, will be explained in another place.

380. Blood serum thus consists principally of *water* holding in solution comparatively small proportions of *albumen and inorganic saline ingredients.*

The oil is not really in solution, and consequently cannot be properly considered a part of the serum. It is merely suspended in the midst of the albuminous materials. The principal ingredients of the blood may therefore be succinctly enumerated in the following list, the numbers expressing the precise value of the relative quantities of each.

Water	795
Inorganic ingredients of a saline nature	9
Crude, plastic material, from which all the organized substance is primarily drawn (*albumen*)	39

(*The above are the ingredients of the serum.*)

Perfected, plastic material, designed for the nourishment of the textures :

Fibrin	2
Hæmatin (including globulin and cell-film)	149

(*These are the ingredients of the clot.*)

Oleaginous matters, partly designed to serve as fuel, and partly to be used in the construction of organs, and mingling both with the serum and clot .	2

The blood, besides the ingredients thus enumerated, contains also very small quantities of certain other substances, which are, however, waste bodies that are removed from it as quickly as possible, and not essential principles intended for use. These will be again alluded to upon a subsequent occasion.

CHAPTER XIV.

THE CIRCULATION.

381. The blood is liquid nourishment, prepared for distribution to *all parts of the living
frame.*

It has been seen that the blood contains within itself
all the nutritious principles of the food, digested and
mixed up with water into a sort of emulsion, that can be
readily forced through tubes, as water is forced through
pipes.

382. The liquid blood is poured through all
parts of the living frame, *by means of a set
of branching vessels* laid down for its conveyance.

Blood-vessels are distributed
from a reservoir and main
branches to all parts of the animal
body, as water-pipes are distributed from a reservoir and main
tubes to all the houses and rooms
of a town. The blood-vessels
branch out into smaller and
smaller twigs, just as water-pipes
are led out from the main in
smaller and smaller tubes. The
figure (*Fig.* 30) represents a portion of one of the small
branches of the blood-vessels, and shows the way in which
these are distributed into twigs. The blood that passes

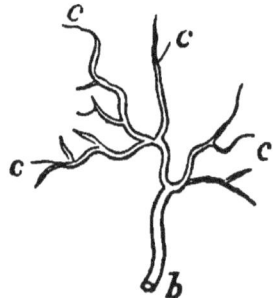

Fig. 80.

through the beginning of the branch at *b*, is scattered to the several minute divisions sketched at *c c c c*.

383. The central reservoir, from which the liquid blood is poured into the branching tubes, is called *the heart.*

The word "*heart*" is derived from the old Anglo-Saxon term "*heorte.*"

384. The heart is a hollow *bag of flesh*, which continues to contract and dilate alternately, so long as it is alive.

When the heart contracts, it squeezes all the blood it contains out through its mouth or orifice. When it dilates, blood again flows in until the interior cavity is filled.

385. The mouth of the heart is *continuous with the great trunk* of the blood-vessels.

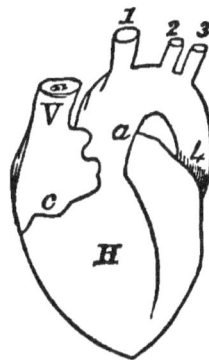

Hence, when the heart contracts, it squeezes the blood that was in its interior forward into the beginning of the branching tubes, that are designed for the conveyance of the stream to the different parts of the frame. The accompanying diagram (*Fig.* 31) shows how the great pipe or blood-vessel comes out from the fleshy bag the heart. H is the heart; *a* is the beginning of the great blood-vessel; 1 2 3 are branches that go upwards from it to supply the arms and head; 4 is the continuation of the vessel running downwards to supply the other parts of the body, and the legs. All these branches divide more and more, and get smaller and smaller, until at last they are like the twigs sketched at *c*, in *Fig.* 30. When the body of the heart at H contracts, blood is forced from its cavity through all the vessels that are continuations of 1, 2, 3, and 4, just as water is forced through the pipes that run from the nozzle of a force-pump, when its handle is worked. The heart is, indeed, a living force-pump, that pumps out blood in the place of water, at every stroke.

386. When the heart dilates after its contraction, the blood, that has just been forced from it, is *shut out from returning* into its cavity by the closing of a valve.

If there were no valve placed at the opening of the great blood-vessel, the blood would flow back as soon as the body of the heart was once again enlarged to its original dimensions. This result, however, is prevented from happening, by the fixing of a valve in such a posi-

Fig. 82.

tion that it allows the blood to pass freely from the heart into the vessel, but altogether prevents it from travelling the opposite way. The figure (*Fig.* 32) represents the arrangement of this valve. *a* is the commencement of the great vessel (*a* of the previous figure) cut open so that its inside may be seen. *p p p* are three little pouches of loose membrane that are pressed close against the side of the vessel when the blood moves from II to *a*, but that are filled and bulged out by it, when it attempts to move from *a* to II, until their edges all meet in the middle, and so entirely close the cavity of the vessel.

387. When the heart dilates after its contraction, blood flows into its enlarged cavity *through a different opening* from that by which it issued on the contraction.

Another large opening admits blood into the heart. This opening is represented in *Figure* 31, at V. In strict accuracy, the cavity of the heart is subdivided into two distinct chambers, which are separated by a valve (at c, *Fig.* 31) that allows the blood to flow from V to II, but prevents it from flowing back from II to V, when the heart again contracts. The chamber which receives the blood from V, is called the auricular chamber. (The word is derived from *auricula*, Latin for "the external ear," which the organ is supposed to resemble). The chamber, which sends forth the blood to *a*, is called the *ventricular* chamber. (The word is derived from *ventriculus*, a Latin word which signifies "a stomach," an organ that this part again has been conceived to be something like).

388. The heart is placed *in the midst of the chest.*

The chest is the upper half of the cavity of the body, divided from the lower half, the abdomen, by the arched diaphragm (299, 703). In this cavity, the heart is carefully suspended, with its bottom pointing a little towards the left side. When the heart contracts, its point is tilted up for the instant, and caused to strike against the outer wall of the chest; the impulse may be felt there by the hand, and is called "the beat of the heart."

389. The vessels, which receive the blood from the heart when it contracts, are termed *arteries.*

The arteries are always found to be empty when examined after death; hence, the old anatomists who first observed them, thought they were only designed to convey invisible spirit or air. They consequently called them by a name that expressed this belief. ("*Artery*" is derived from two Greek words, "*aer*," "*spirit*" or "*air*," and "*terein*," "*to keep*.")

390. The main trunk of the arteries, that first issues from the heart, is termed the *aorta.*

The word "aorta" is derived from the Greek term "*aorte*," which signifies a great artery. The aorta arches over backwards, as soon as it issues from the heart, as represented in *Figure* 31, (giving off branches from its upper part, which run to the head and arms,) and then descends through the back part of the chest and abdomen, just in front of the back-bone, until it finally separates

into two divisions, which pass one to each leg. Throughout this entire course, branch after branch is sent out
from its sides, for the supply of the several neighboring
parts. All these branches divide and subdivide, until
they become small twigs, like those sketched in *Figure*
30. In this way, then, it is that all parts of the body
are supplied with nutritious blood. Blood-vessels branch
out from the main trunk of the reservoir, and, through
these branches, the blood is pumped by successive strokes
of the heart.

391. The twigs of the arteries *terminate in
yet smaller vessels*, which are spread out in
all the structures of the body, like the meshes
of a net.

To the naked eye, arteries seem to terminate as
shown in *Figure* 30, but
when the microscope is employed, it is found that this
is not the real state of the
case. It is then seen that
they end in a mesh-work of
vessels, which are so small
as to be invisible to the unaided eye. *Figure* 33, represents the appearance and arrangement of the capillary
vessels in which the arteries
terminate, when magnified 22
thousand times (150 diameters). *a* is what appears to the
naked eye to be the termination of one of the arterial
twigs, sketched at *c* in *Fig.* 30.

Fig. 33.

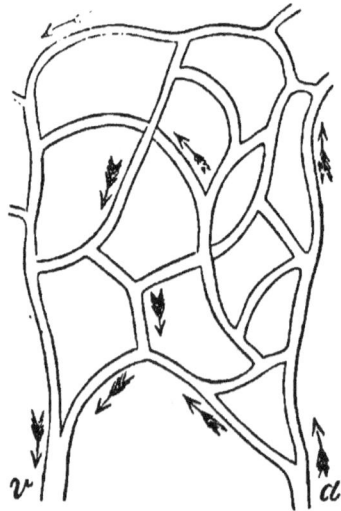

392. The microscopically minute vessels, in
which the arteries end, are termed the *capillaries.*

The word " capillary" is derived from the Latin term

("*capillus*"—*capitis pilus*) which signifies "*a hair.*"
These vessels are so called because they are considered
to be as fine as hairs, but they are literally many times
smaller. The capillary vessels are so minute, that there
is but just room for the red corpuscles of the blood (339)
to be forced through them. As many as three thousand
might be placed side by side within the extent of an
inch, if they were packed closely together, instead of
being looped about.

393. The capillary vessel$ are of *uniform
size* every where.

The arteries get smaller and smaller as they give off
branches, but the capillary vessels do not. They open
freely into each other, still retaining the same dimensions. They intermingle like the meshings of a net, instead of dividing like the branches of a tree. This is
shown in *Figure* 33. The true capillaries are distinguished from the small arteries by this uniformity of size.

394. The capillary vessels are formed of
very *thin and delicate membrane.*

The coats of the arteries are comparatively thick and
strong, but the substance of which the capillary vessels
are formed, is almost as fine and delicate as cell film
(158 and *seq.*). Hence, liquids are able to pass through
them, under the influence of osmose (163) just as they
do through cell films. All the nutritious matters that
get out of the blood to build up structure, or to perform other vital offices, really pass through the delicate
walls of the capillary vessels, under the influence of
osmose.

395. *Every part* of the living body is filled
with the meshes of capillary vessels.

The point of the finest needle cannot be thrust into
any part of the living texture, without wounding some
of them, and allowing their blood to escape. The cavity
of the heart is continuous with the cavities of capillary
vessels that are looped and meshed about in every fibre

11

and membrane of the frame. Hence, whenever the heart contracts, the crimson blood is poured in finely divided floods through all the structures of the body.

396. The blood moves through the capillary vessels with a *steady and continuous* motion.

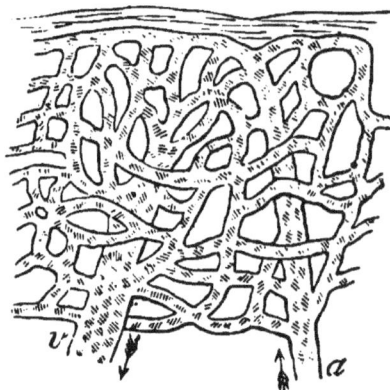

The motion of the blood through the capillary vessels can be seen, whenever any transparent web of a living animal, as, for instance, that which is placed between the toes of the frog, is examined beneath the mi-
croscope, because the
blood corpuscles are
then visible bodies. The
corpuscles are carried
along in the transparent
streams of the blood
plasma, and may be ob-
served making their way
through the twisting
channels, with a steady,
unintermitting, onward
movement. The appear-
ance the blood-globules present to the microscope, while coursing through the transparent capillary vessels of the frog's foot, is shown in *Figure* 34. The little corpuscles may in places be seen to lengthen themselves, as they are pushed by the stream of blood plasma through the narrow channels.

Fig. 34.

397. The *continuous movement* of the blood through the capillaries is caused by the elasticity of the arteries.

The action of the heart is intermitting. This organ contracts and dilates alternately, so that the blood is sent out from it in successive gushes. The blood does not, however, move through the capillary vessels, in like manner, by successive gushes. It runs through them with an even stream. The reason for this is, that the great arteries are elastic, as they would be if made of india-rubber. When the heart contracts, the blood is squeezed out into the trunk of the arteries (the aorta, 390) ; but this and its continuation-branches being elastic, yield to the force for an instant and swell out into larger dimensions. So soon as the heart's contraction stops, however, no more blood is thrown into the great vessel ; it then recovers its size, in virtue of its own elasticity, contracts in turn upon the blood it contains, and, as this cannot get back into the heart (386), forces it on steadily, with a sort of spring, into the capillaries. By the time the artery has recovered its original size, the heart has commenced another contraction; so, in this way, be-tween them they keep the blood moving steadily and continuously on. The beating of the pulse is merely the swelling of the artery when the fresh blood is pumped into it. It is on account of this elastic reaction of the arteries that they are always emptied of blood on the instant of death. (389).

398. The blood does not *move so quickly* in the capillary vessels, as it does in the ar-teries.

This is because the capillary vessels of the body, taken altogether, are of much larger area, than the ar-terial trunk that issues from the heart. Hence, the blood flows as a river does, when it goes from a narrow into a wider channel, that is, with diminished speed. It has been estimated that the entire capillaries of the body have, when taken collectively, an area 400 times larger than that of the great trunk blood-vessel. The blood does not move through the capillaries with a greater speed than about an inch in a minute. Under the mi-croscope, it seems to rush along at a much greater rate, but this is because the space through which it is seen to move, is greatly magnified. When the object is enlarged 150 diameters, the blood that is really moving through only an inch in a minute, seems to be going through twelve feet and a half in the same time.

399. The slow movement of the blood through the capillary vessels favors the passage of its several ingredients *out through their walls*, on the business of nutrition, and for other vital offices.

It is clear that the various principles of the blood are more liable to be drawn through the walls of the capillary vessels by osmose, when they are loitering in contact with them, than they would be if rushing quickly through their channels. There is always a thin layer of blood along the sides of the vessels, that is almost absolutely stagnant and at rest, and the greatest movement takes place in the middle of their channels. It is from this stagnant layer that the matters, used in the work of the body, are principally derived.

400. The capillaries *lead to vessels* that grow into larger and larger branches, by uniting together successively.

Only a small portion of the blood, contained in the capillary vessels at any one time, escapes through their walls by osmose (394). The rest must therefore be pushed on somewhere, or all movement would be arrested, and stagnation ensue. The moving blood is poured from the capillaries into a series of tubes, which are the counterpart of the arteries : that is to say, which unite together to form larger branches, instead of subdividing into smaller ones. The commencement of one of these receiving branches is represented at *v*, in *Figure* 34, and the arrows point out the direction in which the blood flows through the capillary net work from *a* to *v*. The capillary have no other openings than these which connect them with the ends of the arteries, and of those other larger tubes. Their walls are, like cell films, destitute of any apparent pores (160).

401. The vessels that receive the blood from the capillaries, are called veins.

The veins generally remain full of clotted blood after

death. They do not empty themselves at the moment of expiration, like the arteries (389). Hence they appear, in dead bodies, like branching, fibrous cords. They receive their name from this appearance. " *Vein*" is derived from the Greek word for fibre (*is—ina*).

402. The veins are gathered together at last into *two large trunks.*

One of these venous trunks comes from the head and arms, and corresponds with the arterial branches, marked 1, 2, 3, in *Figure* 31. The other comes from the lower part of the body, and accompanies the aorta in its course along the front of the back-bone.

403. The large terminal trunks of the veins *end in the heart.*

The blood that is issued from the heart by the arteries, is brought back to it by the veins. Its course is thus really a *circulation.* It runs in a circle, returning to the spot from which it started. The trunks of the arteries and veins are like twin trees, placed side by side, but with the veins of their leaves incorporated together, instead of being stretched out apart. If such twin trees had only one set of leaves common to the two, and in whose veins the twigs of both were united, it would be possible that fluid might be made to ascend through the stem of one tree, and then be pushed through its branches and through the leaves, and so on into the branches and stem of the other tree, and at last back into the ground, whence it could again start to perform the same journey. In this, the fluid would circle or *circulate* through the trees. This is precisely what happens to the blood-vessels of the animal body. The arteries and veins are twin trunks that branch out together into twigs, which are scattered together through all the regions and structures of the body ; and, between the extremities of each system of twigs, an exquisitely delicate net-work of fine vessels is inserted, to serve as a means of mutual communication. In *Figure* 34, *a* is an arterial branch, and *v* a venous branch ; the rest of the vessels are capillaries.

The arrows point out the direction in which the blood flows.

404. The blood moves at a *slower pace* in the veins, than it does in the arteries.

The veins altogether have three times more capacity in their united channels than the arteries have; hence, as the same quantity of blood has to move through them, it must go three times more slowly. But the veins have a hundred and thirty times less capacity than the capillary vessels taken collectively; therefore the blood moves a hundred and thirty times faster in them than it does in the capillaries.

405. The entire mass of blood, contained in the human body, *circulates* from the heart through the arteries, capillaries, and veins, and back to the heart, *once every minute.*

The heart contracts from 70 to 80 times every minute. At each contraction it sends forth about three ounces of blood; therefore, in one minute, it sends forth from 13 to 15 pounds of blood, which is not far from the entire quantity contained in men of ordinary stature. This rapidity, with which an entire circulation is made, is not at all incompatible with the slow rate at which the blood makes its way through the capillaries (one inch per minute, 398), for each separate portion of blood travels only through a very small capillary tract in getting from an arterial twig to a venous one; probably never further than one-tenth of an inch. This slow portion of its journey, therefore, it can complete in six seconds, and it is quick enough about the rest, for, in the great arteries, it flows with a velocity of twelve inches per second. The greatest length of journey any one single blood corpuscle has to make, in getting completely round the circulation in the human body, is ten feet; if, therefore, it spends six seconds in loitering on the capillary portion of its course, it still has fifty-six seconds for the accomplishment of the rest of its progress. It must be remembered, however, that some portions of the blood pass back to the heart along a much shorter path than other portions. The statements given above are all average estimates. The circulation is really performed through what may be represented as a series of arches of all possible sizes, between half-a-dozen inches and ten feet. The force the heart has to exert, at each stroke, to send its three ounces of blood through this circular course, is equal to a pressure of about 13 pounds.

406. The blood is kept *moving towards the heart* in the veins, by the instrumentality of a series of little valves.

All the veins, that are subject to pressure and inter ference from without, have small valves in their interiors, which are so hung that they lie near to the sides of the vessel, out of the way, when the blood flows towards the heart; but are bulged out, and entirely close the channel, when it attempts to run in the opposite direction. *a*, in *Figure* 35, represents a vein cut open and laid flat, to show its pair of pouch-like valves, lying loosely against the walls. *b*, on the other hand, is the sketch of an unflattened vein, with its valves closed by an attempted backward flow of the blood.

Fig. 85.

a   *b*

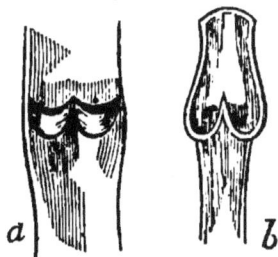

407. The blood, which is conveyed to the capillaries by the arteries, is of a *bright scarlet color*.

Arterial blood is known by the brightness of its color.

408. The blood, which is conveyed back from the capillaries by the veins, is of a dull *purple color*.

Venous blood is known by the dullness of its color.

409. Arterial blood is changed into venous blood *in the capillary vessels*.

The blood goes into the capillary vessels of a bright scarlet color, but comes out of them of a dull purple color. Certain changes are effected in its composition, while passing through these delicate channels, which produce this alteration of hue.

410. When arterial blood is changed into venous blood, it *parts with a certain proportion of redundant oxygen,* which it contained in its scarlet state.

Arterial blood carries pure oxygen gas to all the structures of the body, as well as nourishing material. All the destructive changes, which are effected in the animal organization for the production of power (258), depend upon the union of corrosive oxygen with the various elements of the tissues. Oxygen corrodes the organs of animal bodies, just as it rusts iron, and consumes fuel (35). This is the reason why animal bodies always imbibe oxygen (254) instead of exhaling it, like vegetables. They need its energetic influence for the conversion of the complex principles of which their organs are composed, into the simpler compounds, in order that power may be set free and made available during the conversion.

411. The redundant oxygen of arterial blood is contained in the *red corpuscles.*

These little living cells are carriers of oxygen to all the various structures of the body, as well as of hæmatin to the muscular and nervous textures. They, therefore, call into activity all the changes of composition on which animal life depends, as well as convey the material of nourishment that repairs the waste consequent on the change.

412. The bright scarlet color of arterial blood is dependent upon the *condition of its red corpuscles.* .

It will be remembered that the color of the blood is entirely dependent upon its corpuscles (335). In consequence of the presence of redundant oxygen, the corpuscles of arterial blood are peculiarly dense and firm. Their films are thicker and whiter, than they are after they have lost their redundant oxygen, and reflect more light. The hue of arterial blood is therefore made up of the red color of the hæmatin and the white tint of the cell film, mingled together. The red and the white are blended into scarlet.

413. When arterial blood is changed into venous blood, *waste materials are mingled with it,* **from the various structures and organs.**

As the activity of the vital organs depends upon the corrosion of their organized substance, by the chemical power of oxygen (410), the products of this corrosion must be carried away as fast as they are produced, otherwise they would soon accumulate to such an extent, as to interfere with the performance of the ordinary actions of the economy. The products of the corrosion of the organized textures are mixed with the blood in the capillary vessels, and are then poured off with it through the veins, as waste.

414. Redundant carbonic acid is *mingled with venous blood.*

Whenever complex bodies containing carbon are destroyed, by the corrosive influence of oxygen, carbonic acid must be among the other products of the destructive change (108). All the textures of the body contain carbon. Therefore carbonic acid is abundantly produced in the neighborhood of the capillaries, and is mingled with the blood contained in them. The products of destructive change are conveyed into the channels of the capillaries, by the same influence that the nutritious principles are carried out from them ; that is, by their imbibition through the delicate filmy walls, under the operation of osmose.

415. Venous blood owes its dull purple color to the *condition of its corpuscles.*

The corpuscles of the venous blood have thinner and more transparent films, than those of arterial blood. Hence, the color of the hæmatin is seen through them, without any brightening from admixture with white, reflected light (412).

416. Purple venous blood is *impure blood.*

It contains various waste materials that are no longer of use to the organs from which they have been dismissed. Hence, it is, to this extent, unfit for the purposes of life. It must get rid of its redundant carbonic acid and other waste matters, before it can be again employed in the nutritious service of the system.

417. Impure venous blood is not mixed with the *pure arterial blood, in the heart.*

If it were so, this too would be rendered impure by the admixture, and unfit for the purpose on which it is just about to be sent through the arteries. There are two separate chambers in the heart, set apart for the reception of the impure venous blood.

418. The venous blood is *returned to the right side* of the heart, and the arterial blood is sent out from its *left side.*

There are thus really two distinct hearts placed closely together, the one employed in the service of the pure arterial blood, and the other in the service of the impure venous blood. The aorta (390) goes out from one of these, and the large veins (403) come into the other.

419. The right chambers of the heart *send the venous blood to be purified,* before it is returned into the general circulation for the nourishment of the frame at large.

The right side of the heart contracts and dilates alternately (384), and has two chambers, furnished with appropriate valves, one for the reception of the venous blood from the great vein, and the other for issuing the venous blood, so received, out again in a different direction (387). The right side of the heart pumps the venous blood out from it, by a series of successive throbbings, exactly as the left side does the arterial blood.

420. The venous blood is sent from the heart *to the lungs*, for purification.

The great vessels, which carry the venous blood from the heart, divide and subdivide into smaller and smaller branches, until at last they terminate in a net-work of capillaries, closely resembling those which have been already described (391 and *seq.*). These are contained in the organs termed the lungs, and are called the *Pulmonary capillaries*, to distinguish them from those of the general frame work. ("*Pulmonary*" is derived from "*pulmo*," the Latin for "*lung.*") The blood, after passing through the pulmonary capillaries, enters a series of vessels that are gathered into larger and larger trunks, and that at length terminate in the heart. There is thus a pulmonary *circulation* as well as a general one. The blood circulates from the heart through the general frame of the body, and back, to nourish the structures; and it circulates from the heart through the lungs, and back, to purify itself.

421. Venous blood is changed into arterial blood, in the capillaries of the lungs.

The blood is of a dull purple color when it goes into the pulmonary capillaries, but it comes out from them of a bright scarlet color.

422. When venous blood is changed into arterial blood, in the pulmonary capillaries it is *deprived of its redundant carbonic acid*, and *a redundancy of oxygen is furnished* to it.

The carbonic acid is thrown out into the air with the breath, and the oxygen is taken from the air. Hence, it is, that animals exhale carbonic acid, and consume oxygen (254), and so compensate for the influence of living plants. The nature of the arrangements, by which these changes are effected, will be fully explained in another place. (Chap. xxiii.)

423. The blood returned from the lungs is poured *into the left chambers* of the heart.

The blood returned from the lungs is purified blood, fully charged with oxygen, and ready to be started afresh on its business of nutrition. Hence, it is poured into that side of the heart which has the charge of pumping out the arterial blood. The appearance of the several trunks, which are concerned in the work of carrying arterial and venous blood, from and to the double heart, is roughly shown in *Fig.* 36—which represents the relative situation of the several parts of the organ in Man, *seen from behind.* L is the left side, which pumps out scarlet blood to the system. R is the right side, which pumps out purple blood to the lungs. *s s s s*, are the openings which receive the scarlet or arterialized blood from the lungs, and A is the great vessel through which the arterialized blood then issues to the body at large. V is one of the openings which receive the purple, or impure venous blood from the body at large, and *p p* are the vessels through which the venous blood issues on its way to the purifying lungs. The furrow from *f* to *f* marks the division of each side of the heart into its two chambers, one for receiving and one for emitting (auricle and ventricle, 387). The gen-

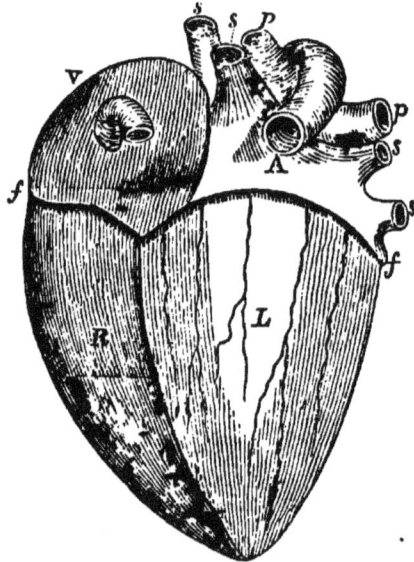

Fig. 36.

eral course of the blood through the double heart, then, is this :—Scarlet blood enters through *s s s s*, gets into *L*, and is pumped thence through the curved vessel *A* to the entire body. Being converted there into purple blood, it comes back to the heart through *V*, and other vessels; gets into *R*, and is thence pumped out to the lungs, through *p p*.

424. There are thus *two distinct circulations* going on in the body at the same time. The heart pumps blood out in both directions at once. Every contraction sends forth three ounces of blood towards the capillary vessels of the general frame, and three other ounces towards the pulmonary capillaries; but every drop of blood of the body, having passed through some of the one set of capillaries, must find its way through some of the other set, before it can return to the first again. Its journey is thus a sort of figure of

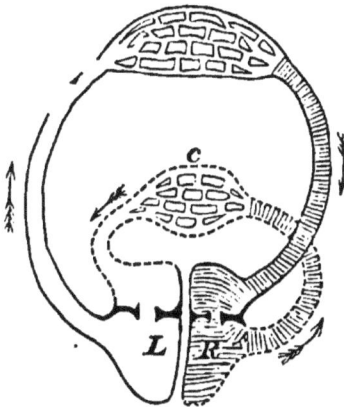

Fig. 37.

8. This may be made very clear by the help of a rough diagram. Let *L* in *Fig.* 37, represent the left side of the heart, and *R* its right side; and let *C* represent the capillary vessels of the body at large, and *c* those of the lungs. Then the continuous outlines show the course of the great circulation *through the body*, and the dotted outlines that of the small circulation through the lungs, the arrows indicating the direction of the movement in either case. The clear parts also particularize the portion of the circulation in which pure arterial blood is moving, and the shaded parts that in which the impure venous blood flows. It will be observed that each circulation is composed half of arterial and half of venous blood, but the relative positions of these halves is reversed in

the two cases. In the great circulation, scarlet blood comes first; in the small circulation, purple blood. In the great circulation, pure scarlet blood is being sent out for the nutrition of the system. In the small circulation, impure purple blood is being emitted for purification. It will also be noticed that the left side of the heart has to do only with scarlet, and the right only with purple blood.

425. The object of the circulation of the blood is thus threefold:—1. *The supply of nourishment.* 2. *The supply of oxygen.* 3. *The removal of waste.*

The circulating blood carries plastic material to vital organs that are exhausting their structures by work (381). It also conveys oxygen to them, to set the exhausting action in operation (410); and it removes, in its backward stream, the refuse and waste produced by the exhausting activity (413). How the waste material is removed from the purple blood will be explained in another place. (chap. xxvi.)

CHAPTER XV.

THE ORGANIC FABRICS OF THE ANIMAL BODY.

426. The plastic materials of the blood *are converted into the fabrics* of which organs are made.

The blood contains in itself all the elements of organized structures. Different kinds of structures select from it, as it flows through them in its delicate, meshing channels, exactly the materials they need for the renovation or extension of their own substances. Bone takes from it the elements of bone; flesh, the elements of flesh; and skin, the elements of skin. A "fabric" is a thing woven or constructed out of different elements. (The word is derived from the Latin term "*fabrico*," to "*make*" or "*build.*") Woollen threads are woven into cloth, and cloth is the "fabric" of which clothes are made. In the same way, "fabrics" are formed out of the elements of the organizable principles of the blood, and organs or instruments are then fashioned from them.

427. *Different kinds* of fabrics are formed for different purposes.

The animal body is a very complex creation, designed for the performance of a great variety of offices. It is consequently furnished with several different kinds of instruments adapted to the diversity of work that is to be done. The plastic materials supplied by the blood are moulded in various ways to make them fit for the construction of these instruments.

428. All the fabrics of the body are formed, either *from living cells*, or through the influence of living cells (249).

It has been seen that the organizable principles are perfected for the use of animals by cell life. Albumen is made in the living cells of plants (242); fibrin is formed by the colorless corpuscles of the blood (363); and hæmatin, by the red corpuscles (344): but, in addition to this, nearly all the substance of animal organization is absolutely *built up of cells*. The *organizable* substances, that have been made *by* cells, are *organized* by being moulded *into* cells, which are then attached together to form structure, in various ways.

429. Some cells, that are destined for the construction of organs, are *developed into specific forms*.

"Development" implies change of character and form. (The word is derived from the Latin term "*develo*," which means to "*discover*" or "*unveil*." When any thing is developed, it discovers or unveils a new nature or appearance.) Development is a species of growth, but it is altogether distinct from the growth of mere size; it is attended by a progressive advance towards perfection, as well as by a progressive enlargement. Thus, if a man took the roof off his barn, built the walls higher, and then put the roof on again, he would make his barn *larger;* but if, besides doing this, he placed partitions and floors across in the interior of the enlarged structure, and furnished them with doors and windows and other conveniences, he would then *develope* his barn into a house. The pale corpuscles of the blood are developed into red corpuscles, by the removal of their tubercles (346), the thickening and contraction of their films, and the deposit in their interiors of hæmatin (347). As the pale corpuscles are developed into red ones, in order that the rich nourishment of muscles and nerves may be perfected, so are other cells developed into a variety of other forms for a diversity of other purposes.

430. Some of the cells of the body are altered *into fibres.*

The sketches in *Figure* 38 show the passage of a cell into a fibre. First, the cell loses its round form and becomes a little oval; then it is drawn out into points at opposite ends; next the points are lengthened more and more, and the internal liquid is evaporated, until at length there remains only a long narrow dry thread.

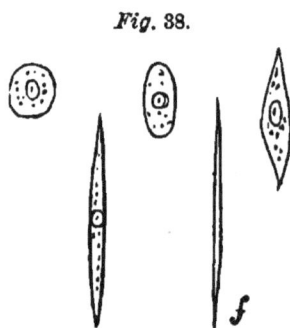

Fig. 38.

431. Perfected fibres possess none of the *vital properties* of cells.

Cells lose all their vitality when they are changed into

fibres. They are killed by the act of conversion. Living cells that are able to multiply, and produce generations like to themselves, always have a trace of a kind of cell germ in their interiors (173). These are seen in the first sketches of *Figure* 38. This trace, is, however, entirely lost in the perfected fibre, shown at *f*. Fibres, hence, cannot produce new fibres, any more than red blood corpuscles can produce new corpuscles. They are *all* originally made from cells.

432. The fibres of the animal body are *woven into tissues.*

The fibres of the animal body are found in the state of connected tissue: that is, they are placed side by side, and made to adhere together ; or, they are interlaced among each other, like the threads of a woven fabric.

433. The animal fabric that is composed of fibres, is called *fibrous tissue.*

A tissue is a thing woven. (The word is derived from the French term "*tisser*," to "*weave.*") Hence, "fibrous tissue" means a fabric of *woven threads* (359). The threads of fibrous tissue are very fine. About twenty thousand of them may be laid side by side, within the extent of an inch. *Figure* 39 represents the appearance of fibrous tissue, seen under a powerful microscope.

Fig. 39.

434. Fibrous tissue is employed for *mechanical* and not for vital purposes.

In this respect, fibrous tissue resembles the woody formations of plants (191). It will be remembered that cells, which are designed for the performance of vital

H*

offices, remain thin and unincumbered by wall deposits
(189).

435. Fibrous tissue is of a comparatively *lasting nature.*

Vital structures are transient, and readily decompose,
because the object of their existence is the extraction of
power out of their destruction (258). But when fabrics
are only made for mechanical service, there is no need
for their speedy destruction. It is desirable, on the
other hand, that they should last as long as possible.
The fibrous tissues of animals undergo very little, and
very slow change, during life. And they also remain
for a comparatively long time undecomposed after death
(188).

436. All the several organs of the body are *connected together*, and *kept in their proper relative positions* by fibrous tissue.

The fibrous tissue, when employed on this service, is
called "*connective tissue.*" It forms a delicate web,
which runs all over the body, so that the different organs may be embedded in it. The connective tissue is
also termed "*areolar,*" because its fibres are so arranged as to leave numerous little void spaces, or interstices (Latin—*areolæ*) between them. The threads of
connective tissue are elastic. They may be lengthened
somewhat, and will then resume their original dimensions when allowed to do so. This is in order that some
relative motion may be permitted between organs that
are connected by means of these woven threads, although
they are prevented from being separated.

437. Fibrous tissue is employed in the formation of *partitions* and *coverings*, in different parts of the frame.

When it is so employed, the fibres are woven into
firm, compact sheets of great strength. Such dense
sheets of fibrous tissue are called "*fibrous membrane.*"
(The word *membrane* is derived from the Latin term

"membrana"—"parchment," "peel," or "skin.") Most
delicate organs in the body are protected by an external
coat, or covering of strong fibrous membrane.

438. Fibrous tissue is used in the construction of *stout cords*, which are employed in
working the movable machinery of the body.
When so used the fibres are placed in bundles, and
fixed firmly together. The cords thus formed are termed
·" *ligaments*" and " *tendons.*" The ligaments bind together the ends of bones, that are connected by hinges
to allow of bending motion. (The word is derived from
the Latin term " *ligo," to bind.*) The tendons unite the
muscles to the bones and to other parts, that are to be
made to change their positions by muscular action. (The
word is derived from the Latin term " *tendo," to bend or
stretch, as a bow is bent or stretched by its cord.*)

439. The vessels of the body (330–382)
are principally formed of *fibrous membrane.*
To form vessels for the conveyance of liquid, fibres
are rolled round and crossed upon each other, until they
constitute layer after layer, and until a sufficient strength
has been secured. In the interior of the fibrous sheath,
a pavement of delicate cells is laid down. There is also
muscular structure mingled with the fibres in the coats
of arteries.

440. The capillary
vessels are exclusively
formed *out of cells.*
When capillary vessels are
formed, simple cells are made
to branch out, and meet by
their extremities, which are
then caused to open into each
other. In *Figure* 40, a cell
is shown in the act of being
converted into a part of a cap.

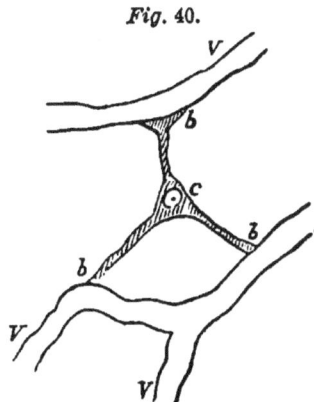
Fig. 40.

illary net work. *c* is the cell containing its cell germ, and *b b b* are three branches extending out from it to unite themselves to *v v* and *v*, capillary vessels previously fully formed. The walls of capillary vessels are as delicate as cell films, in order that liquids may be readily pushed through them by the influence of osmose (394). Consequently these structures cannot be made by building cells or fibres into their walls, as the larger vessels are. Single rows of cells are therefore changed into tubes, their own walls becoming the walls of the tubes, and their own cavities being prolonged into the tubular channels.

441. The digestive organs (281, 299) are principally formed of *fibrous membrane*.

The coats of the swallow, stomach, and bowel, are principally made of fibrous membrane, with muscular structure intermingled amongst its fibres.

442. The digestive organs are lined by *a layer of living cells*.

Some of the living cells, that are in connection with the inner surface of the digestive apparatus, and that have vital operations to perform, have been already alluded to (286–296).

443. The inner surface of the digestive organs, which contains the cell layer, is called "*mucous membrane*."

The cell layer is fixed upon a surface of fibrous membrane. These together constitute what is known as *mucous membrane*. The appearance presented by the pavement of flattened cells, that forms the inner surface of mucous membranes, is shown in *Figure* 41.

Fig. 41.

444. All the cavities of the body, that are in communication with *the external air*, are lined by mucous membranes.

Such passages of the body, as have openings leading to the external surface, are exposed to the chance entrance of foreign bodies, and therefore need some especial provision for protection from the injury that might result from such intrusion. The mucous membranes afford this protection.

445. The mucous membranes *secrete a liquid* from their surfaces, which is called *mucus.*

To "secrete" is to separate. (The word is derived from "*secerno*," the Latin term for *to separate.*) A secretion is a substance separated from the blood. The membranes that line the open passages, separate from the blood a thick, sticky liquid, which serves to protect their surfaces, as a film of gum-water would, and also to wash away light particles of matter that might otherwise remain injuriously attached there. ("*Mucus*" is the Latin word for *a slimy liquid.*)

446. The mucus which covers the surfaces of open passages is *formed by their layers of cells* (441, 442).

The surface of mucous membrane is covered by a layer of living cells, because there are there vital offices to be performed (189). These cells abstract the elements of their slimy secretions from the blood of the capillaries, and then, when they have filled their cavities with this peculiar production, drop off from the surface, burst, and discharge their contents into one continuous stream. The mucous membranes are merely extensions of the outer covering of the body, the skin,—into the internal passages. The skin, therefore, is of a similar character to them; but, in consequence of having peculiar offices of its own to perform, its structure will have to be noticed apart. (Chap. xxiv.)

447. All *secreting organs* are formed of layers or clusters of living cells, lying amidst

capillary blood-vessels, and bound up together in investments of fibrous membrane.

There are several different kinds of secreting organs
in the body. Some separate from the blood matters
that are required for the performance of particular
service in the economy, as, for instance, the saliva (283)
and the gastric juice (286) employed in the digestive process; others separate merely injurious or useless matters,
that could not be retained without causing derangement
to the economy. All of them, however, agree in having
an abundant supply of blood brought to them by capillary vessels to furnish the material of the secretion, and
in containing a vast quantity of living cells adapted to
its separation from that blood. The clusters of secreting
cells, with their apparatus of capillary blood-vessels, are
generally arranged in the form of knotted masses; hence,
they are designated by the name of "*glands*" (349).

448. Secreting glands always have *ducts
for the conveyance away* of their secretions,
as well as blood-vessels for the supply of the
materials contained therein.

Secreting glands are thus little more than modifications of mucous membranes to meet particular needs.
This is very well illustrated in the case of the little tubes
which secrete the gastric juice into the stomach (286,
287). The inner surface of the stomach is generally
covered over by a pavement of mucous-secreting, flattened cells (442); but
at intervals the mucous
surface is sunk into pits,
and in these pits the
covering cells become
bundles of tubes, which
separate gastric juice
from the blood, in the
place of mucus. *Figure
42* illustrates rudely the

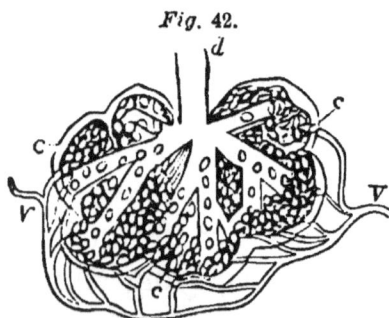

Fig. 42.

general character of secreting glands. *d* is the duct which conveys away the secretion, branching out into the substance of the gland. *c c* are the clusters of secreting cells, and *v v* the blood-vessels which furnish blood for their use. Secreting cells suck the matters they secrete out of the blood (so to speak), and then drop off into the ducts prepared for their reception, like görged leeches, dying and bursting in the act, and so contributing what they had appropriated to the formation of a general outflowing stream.

449. *Fat* is composed of clusters of cells, packed into the meshes of capillary blood-vessels.

Fat is merely solidified oil, stored away in the interior of vesicles, which hang upon the blood-vessels that have furnished the store. Cells that contain fat are very much larger than the other cells of the body.

450. Fat cells are *secreting structures*.

The fixed oil, that fat cells contain, is separated by them from the blood. But fat clusters differ from the ordinary secreting glands in having no ducts through which the separated matter can be poured away. Fat is merely a reservoir of superfluous oil, drawn from the blood and packed away, in order that in any future case of extraordinary need it may be again returned to the blood, to be employed as fuel in the heating service of the frame, and as nourishment for young, forming cells (303, 304). When fat is so returned into the blood, it is again taken up from its store cells, by the same capillary vessels that had served to convey it into its position. Whenever there is a redundant quantity of oil in the blood, it is withdrawn by the fat cells; but it only needs at any time that there should be a deficiency of the same principle, and the action is immediately reversed, and oil is thrown back from the fat cells into the general circulation, instead of being absorbed thence.

451. Fat is used as a kind of soft, elastic cushion *in the package* of the body.

Advantage is taken of the store of superfluous fat being a soft and elastic substance, to turn it to temporary use as a packing material. Nearly all the interstices of the body, that lie between individual organs, are filled with fat. Most delicate structures are padded round with little cushions of it.

452. All the *closed cavities* of the body are lined by *moist, glistening membrane.*

There are numerous closed cavities in the body which have no communication whatever with its external surface, or with external space. These cavities are the hollows in which the several internal organs are packed.

453. The membrane, which lines the closed cavities of the body, is called " *serous membrane.*"

It is so called, because it secretes a thin, clear liquid which is like the serum of the blood, (and which, indeed, seems to be little else). This serous liquid is furnished by a layer of cells, which are spread over the surface of a sheet of fibrous membrane. Serous membrane, therefore, nearly resembles mucous membrane. It differs from it chiefly in forming a thin liquid, instead of a thick one, and in pouring this out upon a closed, instead of upon an open surface.

454. Serous membranes are generally *double.*

That is to say, one fold lines the wall of the cavity, and another covers the surface of the organ that is therein contained. The serous membrane itself is really a sort of bag, with its two sides only separated by the thin, serous fluid poured out upon the inner surface. A rough diagram will represent this most conveniently. In *Figure* 43, let c be the wall of some cavity of the body, and o an organ contained therein, then the double line between them shows

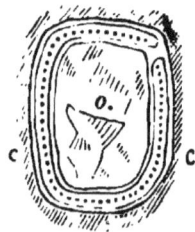

Fig. 43.

the position of the double folds of the serous membrane, the row of dots being in the interior of the serous bag, and representing the fluid with which its surface is there moistened. The one fold of the serous membrane adheres to the wall of the cavity, and the other to the surface of the contained organ, but the moistened surfaces of the two folds are free to move about upon each other. The object of the serous membrane is thus to allow a certain degree of movement in contiguous surfaces without injury resulting therefrom. The serum serves as a sort of lubricating oil.

455. Bone is formed primarily of *living* cells.

Fig. 44.

When bone is first formed, it is of a soft, gristly nature. It is then composed of cells, imbedded in a kind of horny matter, which is produced by themselves. *Figure* 44 shows the appearance of the young soft cells, from which bone is originally formed. ("*Bone*" is derived from the old Anglo-Saxon word "*ban.*")

456. As the structure of bone is matured, the soft cells get *opened into each other*, and become filled with a hard, earthy substance.

Figure 45 shows at *a* the appearance of the cavities when several cell spaces have combined into one, and their walls have been thickened by a hard deposit. In places the cavities get at length completely choked up by hard deposit. Then the bones be-

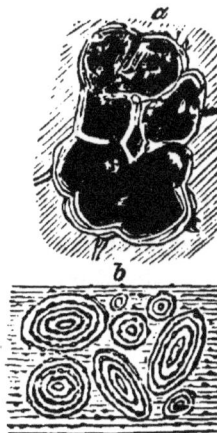

Fig. 46.

come dense, solid masses, capable of retaining permanent forms and resisting very considerable degrees of force when applied to them. The sketch at *b* is a representation of the structure of bony substance, when fully consolidated by its earthy deposits.

457. **The hard deposit, that confers its density and strength upon bone, is principally a *salt of lime*.**

Bone consists of one-third of organized matter, principally of a fibrous nature, and two-thirds of inorganic matter, which is simply grains of phosphate and carbonate of lime (149). All this inorganic matter is, of course, originally derived from the blood. When bones are burnt in a fierce fire, all their organic portions are destroyed, and their earthy parts are left unchanged. When, on the other hand, they are placed in strong acids, the inorganic part is dissolved, and the organic portion left free.

458. **Bone is of a very *lasting nature*.**

Bone lasts for years after the death of the animal, to whose body it has belonged. The fibrous organic part is slowly decomposed, but the earthy portion still retains its original form, until it is finally ground down, rather than decomposed. It is even more enduring than wood (188). Like all other structures of low vitality, the internal changes of bone are effected very gradually, during life.

459. **Bony substance is employed wherever *rigid and unbending* structures are required in the body.**

The bones are moulded into a rigid frame, which is called the skeleton (from the Greek "*skeletos*," "*dry*" —*what remains when the body is dried*). This serves to maintain the general form of the body, to afford points of attachment for the soft textures, and to constitute hard, strong cases for the protection of the most delicate and important organs. The teeth are pieces of bone

covered externally by a coating which has even more inorganic matter in it, and is therefore harder. This hard
coating is called ivory and enamel. The latter has in
itself 97 per cent. of mineral ingredients.

460. Separate pieces of bone are *hinged
together* to form jointed levers, or limbs.

These jointed levers are designed to be worked in the
various locomotive offices of the economy, as will be
more particularly explained in another place (chap. xvi.)

461. The bony pieces of the limbs are
tipped with elastic cushions, where they are
jointed together.

These elastic cushions are formed out of the same
primary cell substance as the perfect bone (455); but
scarcely any earthy matter is deposited in them as they
grow mature. Instead of filling themselves with this
dense substance, they imbed themselves in an elastic
structure which they pour out externally. This elastic
substance pushes the cells further and further asunder,
and compresses them more and more as it is increased.
It is remarkable as being the glutinous principle that is

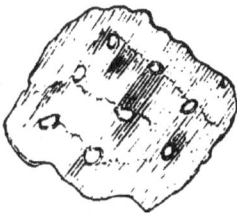
Fig. 46.

familiarly known as jelly, when
dissolved by means of boiling
water. It differs principally
from fibrin in containing less
sulphur and no phosphorus, and
in readily forming chemical compounds with the salts of lime.
Figure 46 shows the appearance
of the substance of these elastic cushions, with their imbedded cells, when mature.

462. The substance of the elastic cushions,
which are attached to the ends of jointed
bones, is called *gristle*, or *cartilage*.

("*Cartilage*" is derived from "*cartilago*," the Latin
for "*gristle*.") Gristle is so soft that it can be cut
with a sharp knife; but it is, nevertheless, very dense

and firm. Its great peculiarity is its high elasticity; it readily yields when twisted or pressed, but immediately recovers its previous shape, when the compressing force is removed. Besides being employed to make the ends of jointed bones elastic, cartilage is always used when orifices or tubes are to be kept open. The external orifice of the ear is an instance of its employment in the one way; and the wind-pipe, which is formed of a series of gristly rings, in the other.

463. The joints are *lined by serous membranes.* (452).

When the ends of the jointed bones have been tipped with elastic cartilage, this cartilage on each side of the joint is covered with serous membrane, exuding its serum so that it is placed like oil between the contiguous moving surfaces. The jointed bones bend and work as if they had a moist, empty bladder between their ends, this bladder being the little sack of serous membrane. The jointed bones are also bound together round their sides by fibrous ligaments (438). The joints are thus made up of bones, cartilage, ligaments, and serous linings.

464. The various fabrics of the body grow and renew themselves by *multiplying their cells.*

The blood plasma is conveyed to all the various fabrics by the capillary vessels, which are meshed about in or close to them. This plasma is pushed by osmose (163) through the delicate walls of the capillaries, and through the cell films, and is then converted into new cell structure, taking the exact impress of the peculiar kind of vesicle under whose influence it is formed. Each variety of cell selects its own appropriate materials from the blood plasma, and makes its offspring like to itself.

465. Some of the cells thus called into existence to extend or to renovate organs, *retain*

their primitive forms and their reproductive powers.

The organs made through their instrumentality are then vital organs, and undergo rapid changes of internal composition, which changes are all connected with the performance of the functions of cell life (180). All the modifications of the secreting organs are illustrations of structures of this kind, that possess high vitality.

466. Others of the cells, so called into being, lose their primitive forms, and their powers of reproduction, and become *developed or changed* into new states, suitable for moulding into less fragile textures.

The organs made by their instrumentality are then mechanical, and not vital organs, and they undergo little further change of internal composition after they have once been completely formed. The bones, gristle, and fibrous structures generally are illustrations of mechanical organs of low vitality.

467. All the fabrics enumerated above are the materials which are employed in constructing either apparatus that does *organizing work*, or *instruments* that are for *mechanical* purposes.

They are merely fabrics, which are made subservient to the support of the higher organs, that are the especial seat of animal life, as contrasted with organic life (271, 372). The digestive apparatus prepares the blood, the blood-vessels distribute it, the secreting glands furnish liquids that are needed in the digestive process, and remove waste material from the blood, in order that it may be kept pure ; and so, on the other hand, the bony skeleton preserves the treasures entrusted to its caverns, and its levers bend when pulled, or stand rigid when left alone. The serous membranes oil the joints and movable parts. The mucous membranes protect exposed

surfaces, and the fibres bind, connect, and cover. In all
this, however, there is nothing of voluntary power, feeling,
or perception. Animal life is a something superadded
to all that has been hitherto spoken of, and has especial
fabrics of its own.

CHAPTER XVI.

THE MUSCULAR APPARATUS.

468. The power to *move at will* is one of
the characteristics of *animal life* (272).

This power is evidently conferred on the animal in
order that it may vary its relations to external things.
It would have been of no practical use that animated
creatures should have been able to notice and perceive
the objects by which they are surrounded, unless they
had also had the power of altering their relations to
them. It would have been of no use that the grazing
animal could perceive that it had consumed all the grass
within its reach, unless it had been able to start off in
search of fresh pasture. It would have been of no use
that the beast of prey could see its food grazing on the
plain, unless it had also been capable of bounding off in
its pursuit. It would have been of no use for man to
observe and ascertain the conditions and laws of material
nature, unless he could lay his hands upon matter, and
fashion it to his purposes. There are certain movements which take place in vegetable structures; the rise
of the sap, for instance, into the leaves; the growth of
organized substance; the bending of twigs and foliage
before the wind; and even the shifting of small, light
bodies in a mass from place to place (251). All these
movements are, however, to be readily traced to the
operation of external influences. They none of them

originate in the organization, in answer to impressions made from without. The power of originating movements in their own frames, therefore, at once distinguishes animals from plants, and is a marked characteristic of animal life.

469. Animal movements are effected by means of a *special apparatus* added to the body for the purpose.

This apparatus is, therefore, composed of instruments which belong to animal, rather than to organic life. It comprises an important part of the *animal organs*, which are superadded to the nutritive and mechanical organs, for the perfection of the animated frame (467).

470. The organs that are employed in effecting movement in animal structures, are called *muscles*.

(The word "*muscle*" is probably derived from the Hebrew term *muscut*—"*contractility ;* MSC.—"*to contract.*") The flesh of animals is mainly muscular substance.

471. Muscles are composed of *bundles of fibres* bound up together in sheaths of fibrous membrane.

Figure 47 shows the appearance of a small piece of muscle cut across. It is magnified about 25 times. Its division into bundles of fibres is seen at the cut end. Each of these bundles is inclosed in a sheath of fibrous membrane, and in that state is termed a "*fasiculus,*" (the Latin term for a "*packet*" or "*little bundle.*") As the fibres are wrapped up in fibrous membrane to make fasciculi, so the fasciculi are wrapped up in fibrous mem-

Fig. 47.

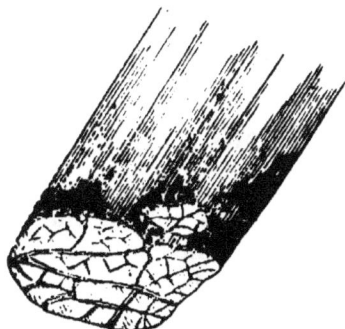

brane to make muscles. Each muscle has its own external sheath or covering. A muscle is a bundle of fasciculi: a *bundle of little bundles*.

472. Each muscular fi-
bre is itself a *bundle of
fibrils*.

Fig. 48.

In *Figure* 48 the appearance
of a portion of muscular fibre is
shown, very highly magnified.
At the upper part the several
fibrils are seen escaping separately from the investing sheath
which forms the covering of the
fibre. (The word fibril is merely
diminutive of fibre. It means a
little fibre.)

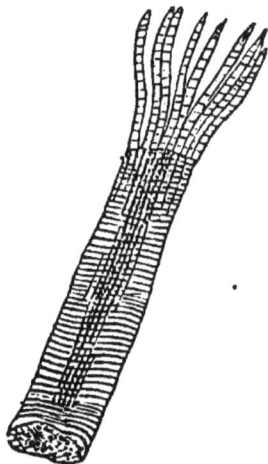

473. The fibrils of mus-
cle are *extremely small*.

Muscular fibres are very coarse compared with those
of the common fibrous tissues (433). They may be
readily distinguished by microscopes that magnify only
a few times. About 400 of them will lie side by side
in the extent of an inch. But each fibre contains
Fig. 49. several hundred *fibrils;* these, therefore, are very
small and require that very powerful microscopes
should be employed before they can be discerned.
Between thirty and forty fibrils may be placed
within the breadth of one fibre, and as many as
fourteen thousand of them in the extent of an inch.

474. Every *fibril* of muscular fibre is
itself a *row of cells*.

In *Figure* 49 the appearance is shown that a
single fibril presents, when magnified above fifty
thousand times. It will be seen that it consists
of a row of extremely delicate cells, placed end to
end. The dark spaces are the cavities of the cells,
and the clear spaces their films or walls. A mus-

cular fibre, therefore, really consists of several rows of cells, piled end to end, and placed together in a common tube. Each cell of the muscular fibril is one-half longer than it is broad.

475. The several cells of the muscular fibril *adhere together by their ends* with considerable force.

Hence, each row really acts as a connected fibre. It is indeed a fibre *built up of several distinct pieces*, mechanically adjusted together, instead of being merely formed as one piece. There is, however, some adhesion present between the cells of the separate rows, for occasionally a fibre has been seen to separate into transverse rows instead of longitudinal ones. The cells then have stuck together by their sides, and separated at their ends.

476. The sheath of the muscular fibre is a *transparent membrane.*

Hence, the markings of the rows of cells may in some places be seen through it. On this account the unopened, muscular fibre seems generally to be marked across by transverse streaks. This appearance is represented in *Figure* 48. The membranous sheath of the fibre is, nevertheless, very strong and tough. It will remain untorn long after the contained rows of cells have been pulled asunder. It is called "*sarcolemma,*" (a word derived from the Greek, *sarkikos*—"fleshy," and *lemma*— "bark").

477. The cell of the muscular fibril possesses *a peculiar power of altering its form,* when stimulated to do so, by the application of a certain influence.

When the cell of the muscular fibril is irritated, as, for instance, by the contact of a sharp point, it all at once contracts its length. It does this, however, by changing its shape. It does not shrink in its entire dimensions, but increases in breadth as much as it diminishes in

13   I

length. It bulges out one way, while it falls in the other.
From being half as long again as it was broad, it becomes all at once as broad as it is long.

478. The peculiar contractile power of the
muscle cell is called its *contractility*.

"Irritability" is another word for the same property.
"Irritability" or "contractility" are the characteristics
of muscular structure.

479. When the cells of a muscular fibril
contract, the *fibril itself is diminished in
length* in the same proportion.

If all the cells contained within the tubular sheath of
a fibre diminished their lengths one-third at the same
time, the fibre itself would of course also become one-
third shorter.

480. When a straight fibril is diminished in
length, its opposite ends are made to *approach
nearer together*.

And, consequently, if two distinct bodies, that are
capable of being moved, are attached to those ends, they
too are brought nearer together. Most of the muscular
movements of the body are effected in this way. Cords
composed of muscular fibres, instead of the more common ones, are stretched between two bodies, one or both
of which are to be moved. These cords are made fast
by their opposite ends to the bodies; and then, when
the cords contract their lengths, the bodies are pulled
together.

481. When muscular fibrils are contracted
in length, they are proportionately *bulged out
at their sides*.

What is true of the cells (477) in regard to their
change of dimensions, is of course also true of the fibrils
that are formed of them; and what is true of the fibrils,
is also true of the fibres formed out of the fibrils; and
of the fasciculi and muscles made by the fibres. Muscles

bulge out when they contract, in exact proportion to the diminution of their lengths.

482. The fibrils of muscles are composed of cells, because they have *vital offices* to perform.

Change of form and dimension, under the application of irritation, is a vital action. The cells of muscular substance retain the perfect cell form, and are not changed into simple fibres, like the cells of fibrous tissue (433), because it is required that they should possess the power of performing this operation.

483. The cells of muscular fibrils *are wasted by their contractile efforts.*

Every time a muscle cell contracts, a portion of its substance is destroyed. This is one illustration, then, of the fact that, in the animal body, power is procured through the destruction of organic material (258). The mere organic actions of the body are constructive, like those of a plant. They dissolve, circulate, and make more plastic the nutritious principles plants primarily form. But the first really *animal* operation we have to consider, the contraction of the muscle cell, is connected with destruction of the fabrics built up out of the nutritious principles.

484. The waste of the muscle cell is *made good from the blood.*

Fig. 50.

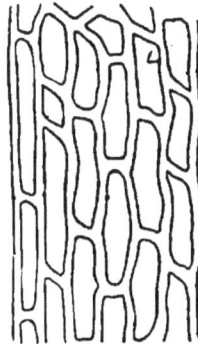

Living muscles are abundantly supplied with blood. Large arterial branches are sent to them, which terminate in capillaries that are looped about in all directions amidst' the bundles of fibres. *Figure* 50 furnishes a sketch of a small portion of the capillary network that is distributed in the midst of the muscular substance.

485. The waste of the muscle cell is effected

through the instrumentality of *corrosive oxygen* (34).

The complex organic substances, of which the cell and its contents are composed, are resolved into carbonic acid, water, and other binary compounds, by oxygen seizing upon their elements (39). Oxygen thus wastes and corrodes the substance of living bodies, as it does most of the other materials of nature (35).

486. Oxygen is *conveyed to the muscle cells* by the blood.

The red corpuscles of arterial blood contain a quantity of superabundant oxygen (411). This oxygen they give up to the muscle cells, when they pass through the capillary vessels that lie amongst the muscular fibres.

487. The blood constituent, that is especially devoted to the support of muscular action, is *the hæmatin* of the blood corpuscles (342).

The red color of muscular substance is an indication of its close connection with the coloring principle of the blood. The blood corpuscles thus carry to the muscles at once the chemical influence that determines their waste, and the nutritious principle that effects their repair. A very highly animalized and finished product is prepared for the support of muscle cells. The ordinary plastic principles of the blood do not seem to be equal to the construction of their irritable substance.

488. The waste, that results from the corrosive union of oxygen with the elements of the muscular substance, is *conveyed away by the veins*.

The carbonic acid and other products of decomposition, that are set free when the organized substances of the muscle cells are disorganized for the production of their contractile efforts, are carried into the delicate capillary vessels, by the influence of osmose, and then pass on from them into the venous branches that form

their continuation. The capillaries of the muscular
bundles, like all the other capillaries of the frame (400,
401) lead to veins.

489. The capillary vessels supplied to the
muscles, do not *penetrate into the sheaths of
the fibres* (476).
They are merely meshed about everywhere between
the fibres. Whatever passes from the capillary vessels
into the muscle cells, has to be pushed by the force of
osmose through the delicate walls of the vessels, the
membranous sheath of the fibre, and the films of the
contained cells. And whatever is returned from the
cells into the capillaries, must permeate in the same
way, through the same textures. Hence, in a general
way, the fineness of the fibres of a muscle and the abundance of its blood supply are proportional, each to the
other. The most energetic and active muscles have the
finest texture, and the freest supply of blood.

490. Muscular fibres are generally attached
at their ends *to tendons and membranes.*
And these tendons or membranes are in turn fixed
into the bodies that are to be moved by the contraction
of the muscular fibres. The vital and contractile part
of the cords, used in working the mechanical apparatus
of the body, passes into uncontractile and simply fibrous
structure, that is only capable of performing the mechanical office of attachment. The fibres of the mechanical
structure are loosened out from each other, and then
fixed round the end of the sheath of the larger muscular
fibre. Muscles nearly always terminate at either ex-
•tremity in tendons, or dense fibrous membranes. The
arrangement of the muscular and tendinous fibres is,
however, varied to a very considerable extent in different
cases, according to the precise nature of the end to be
attained.

491. Muscles *get exhausted by action.*
When muscles act vigorously, a great number of the

cells of their fibrils, and a great quantity of the substance therein contained, are destroyed. Muscles then lose their contractile powers, until there has been time for new cells to be produced, and for new organic substance to be supplied. No muscular action can therefore be continued for a long time without an interval of repose.

492. All the cells in the fibrils of a muscle do not *contract at the same time.*

No individual cell can continue its contraction for a lengthened period. It gets exhausted by the mere act of changing its form, and therefore must be recruited before it can do the same again. It often happens, however, in the mechanical operations of life, that a muscle is required to be kept in the contracted state for some time, as when any heavy body is held up firmly and sustained by the hand. This state of prolonged contraction is effected by the different cells of the muscular fibrils taking up the contractile effort in turn. When a muscle is held contracted, some of its cells are shortened and some lengthened, but at every instant these exchange their condition with each other. Every cell is in a state of repeated and alternated contraction and relaxation. The cell contractions run through the muscular fibrils, as the oscillations of ears of wheat run through a corn field, when the stalks are swayed by the wind. In this way, opportunities are offered to each cell of getting itself repeatedly refreshed for renewed effort, and yet, during the whole time, the contraction of the muscle itself is kept up. When a muscle is sustained in vigorous contraction, the continued and alternate action of its cells may be heard going on as a faint, thrilling sound.

493. Most of the muscles *are placed round the bones* to which they are attached.

It is in this way that the bones are covered with flesh. All the muscles of the limbs are thus packed and arranged. A limb is made up principally of a skeleton of bones, furnished with appropriate joints,—of layer upon layer of muscles and tendons, to move them on each other,—of fibrous membranes for packing them and covering them up,—and of blood-vessels for the supply of oxygen and nourishment to the whole. The arm and hand consist of no less than thirty separate pieces of bone, hinged together by joints, and these are worked by as many as 53 separate muscles. The leg and foot are made of 29 pieces of bone, and they are worked by 56 muscles. Anatomists have given names to nearly 600 distinct muscles in the human frame, each, of course, formed of its myriad fibres, and its thousands or millions of cells.

494. Where two bones are to bend upon each other, the muscles that effect the movement are carried from one to the other, *across the connecting joints.*

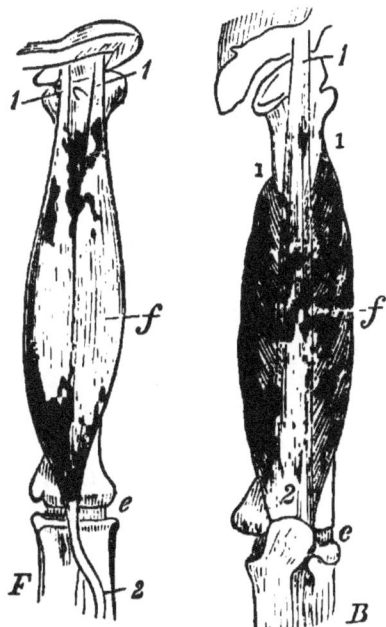

Fig. 51.

The bones of the limbs are pulled backwards and forwards upon their hinges by their contracting cords, as a door may be pulled open, and may be shut, by ropes going opposite ways from its handles. Generally one of the bones is fixed, when the movement is to be made, so that the other may be bent towards it. The sketches in *Fig.* 51, will serve to illustrate the manner in which the movements of the jointed limbs are effected by muscular contraction. *F* represents the front of the arm, and shows the muscle by whose contraction the fore-arm is bent upon the elbow. 1, 1, and 2, are the tendinous attachments of the muscle above and elbow, and f is its fleshy part, lying in front of the upper armbone, e being the elbow-joint: when this muscle contracts the ends 1 and 2 are brought nearer together, and, consequently, as the upper end is fixed to the shoulder at 1 and cannot move, 2 is brought up towards 1,—the forearm is bent at the elbow,—and the hand lifted towards the shoulder. B represents the back of the arm, and shows the muscle by whose contraction the forearm is extended upon the elbow. 1, 1, 1, and 2, are the tendinous attachments of the muscle above and below, and f is the fleshy part, lying behind the upper arm-bone, e being the elbow-joint. When this muscle contracts, the end 2 of the forearm is brought back on the elbow, and the hand carried down from the shoulder. It will be seen that the muscles in A and B are opposite muscles, and antagonize each other. The one undoes what the other has done; when the one is contracted, the other must be relaxed.

495. The body is moved as a whole, from place to place, by making the lower end of the jointed leg *a movable extremity and a fixed prop*, alternately.

In walking, the trunk of the body is first fixed on the right leg, formed of several bony pieces, as represented in *Figure* 52, at L, the undermost of them constituting the arch of the foot. Then, by the contraction of powerful muscles, placed in front of the bones, the foot of the left leg is brought forward and planted on the ground. Next by the action of the several muscles of the right leg, the weight of the body is pulled forwards, and thrown upon the advanced foot, which, in its turn, becomes a fixed pedestal whilst its neighbor is brought forward. The *Figures* A and L show the different arrangements of the bony skeleton in the arm and the leg; the one designed for free and delicate movements upon the shoulder as a fixed point—the other, for the performance of limited movements, requiring great strength and powers of resistance. In the arm (*A*), the upper bone is capable of being swept freely round upon the shoulder-joint (*s*) in all directions. The forearm bends up and down upon the hinge-like elbow (*e*). The large bone of the forearm, and the hand (*r*) roll over the small bone (*u*) backwards and forwards, and the taper pieces of the

Fig. 52.

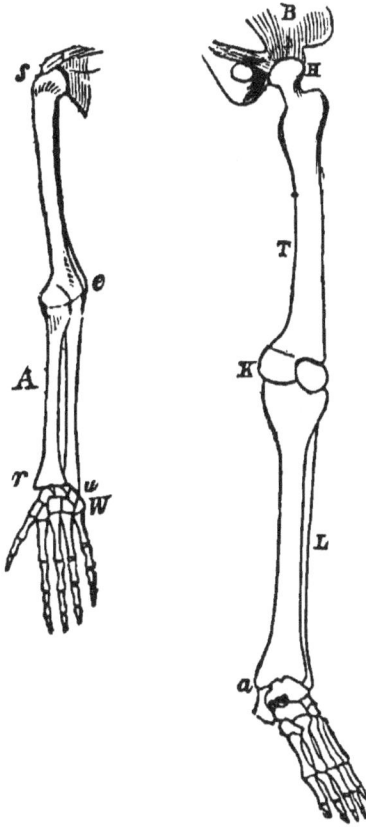

fingers bend backwards and forwards upon themselves and upon the eight bones of the wrist (*W*) by hinge-like joints. All these movements are effected by muscular cords appropriately distributed before and behind. In the leg (*L*) the thighbone (*T*) rolls backwards and forwards, and outwards, at the hip-joint (*H*), which attaches it to the trunk of the body (*B*), and the leg (*L*) bends backwards and forwards upon the thigh at the hinge-like knee-joint (*K*). The seven bones of the ankle (*a*) also allow a certain degree of play between the foot and the leg, in order that their relative positions may be somewhat altered, as the foot is thrown forward, or as the leg is brought over the foot; but the foot itself is modelled into an arch, upon the crown of which the leg is fixed, so that its weight, and that of the body, when brought forward, fall upon a strong and yet elastic base, able at once to support and break the shock of the consequent impulse. It will be observed that different as the uses are, to which the arm and the leg are put,

I*

they are nevertheless moulded upon the same bony elements (with the exception of one bone less in the ankle
than in the wrist, and one more in the knee-joint than
in the elbow). There are in each, the one arm or thigh
bone; the two fore-arm or leg bones; the eight and
seven wrist or ankle bones; the five hand or foot bones;
and the fourteen finger or toe bones.

496. Some muscles are arranged as *broad,
thin layers* round the cavities of the body.

They then serve to vary the size of those cavities, by
their contractions. The walls of the abdomen (299) are
principally composed of layers of muscular fibres and of
strong fibrous membrane, alternating with each other.

497. Some muscles are arranged as circular
fibres *round openings or tubes*, which their
contractions are designed to close, or to make
more narrow.

The lips are chiefly composed of a circular band of
muscular fibres, arranged round the opening of the mouth. *Figure* 53 represents
this arrangement, seen in profile. The
mouth is closed and compressed when the
fibres of this circular muscle are contracted. Muscular fibres are placed
transversely round the tubes of arteries,
between the outer and the inner membranous coat. It is their office to regulate the supply of blood through the particular vessel to which they belong.
When they are contracted, the bore of
the tube is narrowed, and, consequently, less blood is
able to pass through the channel in a given time.

Fig. 53.

498. Some muscles are arranged circularly,
crossing and recrossing each other, so that
they *make up hollow organs*, possessing contractile power.

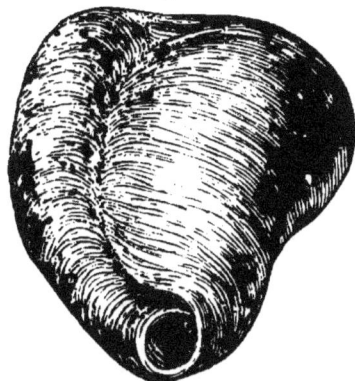

The heart is the most perfect illustration of this kind of muscular arrangement. *Figure* 54 shows how the entire walls of the organ are composed of thick, powerful bands of muscular fibre, distributed circularly. When these bands are contracted, the cavity of the organ is obliterated for the time, and the blood forced from it into the vessels (384). The coats of the swallow, stomach, and bowel, have muscular fibres interwoven amidst their substance. When any of these fibres are contracted, the parts of the membranous coat, to which they are attached, are brought nearer together. The muscular fibres of these organs do not, however, contract all at one time. Some contract first, and then others do the same, upon these relaxing. The contents of the organs are thus moved, from one place to the other within them, instead of being at once expelled, as is the case with the heart.

Fig. 54.

There are thus three principal ways, in which the contractile ropes of the muscular apparatus are arranged so as to produce motion. First by being stretched between two distinct pieces of rigid bone, which are so hinged together as to be able to move upon each other. Secondly, by being placed in the form of continuous rings round openings which require to be closed, or round tubes which need to be occasionally narrowed; and thirdly, by being woven into the walls of hollow bags, which hold matters that need to be expelled, or moved at intervals. There are other methods in which muscular fibres are made to effect movement, as when they are attached to the fibrous covering of the eyeball, in order that they may roll it about within the socket (673), or when they arc fixed to the cartilaginous pieces at the top of the wind-pipe, which form the organ of voice (685). Still, in every case, the principle of muscular action is the same, the bringing towards each other of movable objects, by means of connecting fibres, that are capable of shortening their own lengths.

499. Some muscles *contract and dilate* independently of *all operation of will*.

The heart does this; it contracts and dilates about 70 times every minute through the entire period of life, without stopping, although many people do not even know that they have hearts within their chests. Muscles that contract, in this way, independently of any exercise of will, are called involuntary muscles. They are employed in the performance of operations, whose continued repetition is essential to the support of life, and which, consequently, cannot be intermitted even during sleep. The beating of the heart, and the play of the chest in breathing are of this kind. The fibres of the involuntary muscles are, mostly, of a less perfect kind than those that are destined for the performance of voluntary actions. In many, no separate cells can be distinguished in the fibrils. In the fibres of the heart, the banded appearance, which indicates the presence of these cells, can be discerned ; but it is far less distinct than in the muscles employed in effecting the voluntary movements. It seems as if the involuntary operations were carried on by a series of fibres of a lower order and less perfect development than those which are under the direction of the will.

500. A certain amount of *vigorous exercise* is essential to the formation, and the healthy action of the muscular apparatus.

Muscular structure is built up in proportion to the call there is for its presence. The man, who makes very little use of his arms, gets no strength in them ; but the blacksmith soon has arms that are equal to great exertion. When great waste is produced in muscular substance by its employment, increased quantities of blood flow to it for the repair of the waste; but portions of this blood are then also used in making the muscular substance itself larger. New fibres are formed at the same time that the old worn ones are repaired.

501. *Intervals of rest* are essential for the preservation of the vigor of the muscular apparatus.

The muscles become fatigued from long-continued exertion. This is because, during such exertion, their little cells are destroyed, and their contractile substance is consumed, faster than new vesicles can be formed, and fresh material be supplied. During repose, however, cells are made, and contractile substance is built up, whilst scarcely any waste is experienced; consequently, by it the muscular apparatus is soon rendered fit for renewed exertion. There is no exception to this law, that muscles must rest as well as work. So long as a muscle cell is in action it is wasting; after the action, there must therefore be an interval of rest, in order that the waste may be repaired before it begins again. The provision that has been made under these circumstances for the support of continuous muscular contraction, by causing the little cells to carry on the operation by turns, has been noticed. In muscles that go on acting incessantly through life, like the heart, the action itself is an intermitting one. Fibres are constantly resting as well as constantly acting. Every contraction of the heart is followed by a relaxation. It is during that interval of relaxation, the repair of the organ is carried out, and its wear and tear made good.

502. Muscular exertion *quickens the circulation* of the blood.

It effects this primarily through the instrumentality of the veins. Venous trunks run along amidst almost every large collection of muscular bundles; but when these muscular bundles contract, they of necessity compress the veins lying in contact with them, and squeeze the blood out of their interiors. This blood can, however, only flow in one direction, in consequence of the numerous valves with which the interiors of the veins are furnished (406). It can only run *towards the heart;* but the increased quantity of blood that enters the heart, stimulates it also to augmented activity, and makes its contractions more frequent. Muscular exercise thus acts as a stimulating influence on all the vital organs of the frame. Where the body is vigorous and healthy, it tends to exalt its life; but where it is weak and in disorder, exercise leads at once to further exhaustion, and increases the mischief. Great muscular exertion is a delight, and a healthful enjoyment to the strong; but it is a misery and a danger to the weak.

503. A free *supply of blood* is essential to muscular activity.

The blood acts at once as a stimulus to muscular contractility, and as a supply of strength to the contracting organs. It carries, it will be remembered, to the muscles, both oxygen to set up the corrosive decomposition of the cell substance, out of which the activity is extracted (411), and plastic material for the repair of the decomposed substance (381). In fainting fits, the heart ceases to act for a few instants, and circulation is therefore stopped. At such times, no blood is sent to the muscles, and, consequently, they are weak, and without power until the stimulating and nutritious steam begins to circulate again amidst their fibres.

CHAPTER XVII.

NERVOUS APPARATUS.

504. Some *other influence*, besides free supply of blood, is requisite for the production
of muscular contraction.
Blood is always circulating through the capillaries of
the muscular fibres, but the muscles are not always in
action. The presence of another influence, of the nature
of an irritation or stimulus, is necessary for the production of muscular contraction. Irritation can produce
contraction in the muscles of dead animals, through
which blood no longer circulates. Pressure upon the
muscular fibres by the point of a needle, or the passing
of an electric shock through them, will cause twitches in
the movable parts to which they are attached. This,
however, can only be done for a very short time. The
contractile power of the fibres is soon exhausted by
these applications, and then is not capable of being renewed. The blood, therefore, acts by *keeping up the
contractile power* of muscles. It furnishes the fresh oxygen and nutritious matter which are essential to the continuance of the process; but it does not itself make the
fibres contract.

"*Stimulus*" is the Latin word for a "*goad*" or "*spur*."
A stimulus is an influence that goads or spurs on the
performance of an action. To "irritate" is to "provoke" or "stir up." (It is derived from the Latin word
"*hirrio*," to "snarl," which is what dogs do when they

are provoked. The muscular fibres are irritated or provoked, when they contract, as dogs are when they snarl).

505. The influence which "provokes" muscular fibres to contract, is *brought to them by a quantity of delicate white threads.*

When the fibrous bundles of a muscle are closely examined, it is observed that a quantity of delicate white threads, as well as the capillary blood-vessels, are looped about amongst the fibres. *Figure* 55 shows a series of these white looped threads distributed amidst a few muscular fibres.

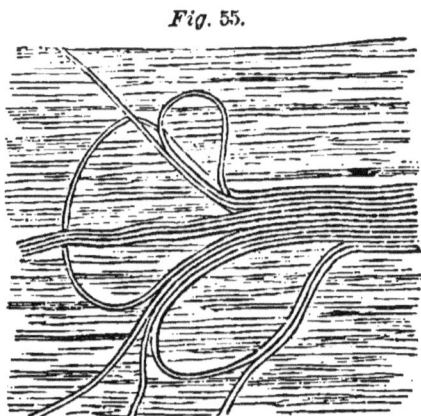
Fig. 55.

506. The white threads, which bring stimulant influence to the muscular fibres, when they contract, are called *nerve-threads*.

The word *nerve* is derived from the Greek term "*neuron*," the "string of a bow." The larger nerves look like glistening strings, when their course in the dead body is exposed. Hence, the first anatomists who observed them, called them "strings," and the name has continued in use to the present time.

507. The nerve-threads, distributed amongst the muscular fibres, are of very *minute dimensions.*

They cannot be seen without the assistance of a powerful microscope. They are smaller by four or five times than the capillary blood-vessels (392). In some places they are so minute, that as many as twelve thousand of them would lie side by side, within the extent of an inch.

Nerves that can be seen by the naked eye, are composed of a great many of these threads bound up together.

508. The nerve-threads are really "*tubules*," or little tubes made of very delicate films.

Fig. 56.

The nerve-threads are formed of delicate, transparent films, moulded into the shape of tubes, which are filled with a white pulpy substance that is seen shining through their thin walls, and that thus confers upon them their glistening appearance. *Figure* 56 is a representation of a portion of three or four nerve-tubules, as seen when magnified many thousand times.

509. The nerve-tubules loop about amidst the fibres of muscular subtance, without, however, *penetrating their sheaths*, or opening out into them.

In this respect the nerve-tubes are like the capillary tubes of the blood-vessels. They are distributed amongst the fibres of the muscles, without any communication being established between the interiors of the two. Whatever influence passes from the nerve-tubule, and reaches the muscle cell, does so, like the blood principles (394), by penetrating through the wall of the tubule, the sheath of the muscular fibre, and the film of the contained muscle cells. The nerve-tubules, like the capillaries, too, have no distinct terminations, they loop round, and return the same way they came; but they do not open into each other, as the capillary vessels do. Each one *keeps its interior channel perfectly distinct from its neighbor's.* This method of distribution is represented in *Figure* 55.

510. The nerve-threads come to the muscles as *large branches*, and are then distributed in all directions.

14

Large branches of nerves enter the muscles, just as large branches of arteries do; and these branches are then subdivided into smaller and smaller threads, which are scattered in all directions, for the distribution of nerve-influence. In *Figure* 55, the twig of a nerve is seen subdividing into its smaller threads, which are separately scattered amidst the muscular fibres.

511. The nerve-branches are *bundles of distinct tubules*, bound up together in a common sheath.

They are not ramifying tubes like the blood-vessels (391), with channels that are subdivided, and continuous with the smaller ones. They are distinct, isolated tubes running together side by side for a certain distance, and then parting company to be severally scattered in different directions. The existence of the several tubules of one nerve-branch,

Fig. 57.

as isolated and separate threads, is marked in *Figure* 55. In *Figure* 57, a magnified sketch is given of a portion of a nerve, cut across to show the separate bundles of tubules lying within it side by side.

512. The nerve-tubules are kept distinct from each other, because they are destined for the conveyance of *distinct messages to particular places*.

If the nerve-tubes branched out like the blood-vessels, they could only pour out the stimulant influence in a general way, as the blood-vessels pour out nutritious blood, and so would set all the muscular fibres of a part convulsively twitching at once. What is needed, however, of them is to cause some of the muscular fibres to contract at one time, and others at another time, according to the particular mechanical work that is to be done. They have to carry particular messages to particular

places. In this respect they are like the wires of the
submarine electric telegraph, which are separated from
each other by sheaths of gutta percha, and are then
bound up all together into a cable, that is laid along the
bed of the sea. At the end of its course, each wire tells
its own tale, although it has travelled in close neighborhood with several companions, as distinctly as if it had
been separated from them all the way by hundreds of
feet. As the great main of a water company, distributed
into the several houses of a town by branching pipes, is
a type of the arrangement of the blood-vessels, the telegraph cable, composed of separated wires bound up together for a certain distance into a common strand for
the convenience of transmission, and then sent off, one
to the east, and another to the south, is a type of the
nerve arrangement. The blood-vessels distribute nutritious blood *in all directions*, as water pipes distribute
water. But the nerves send distinct messages in *particular directions*, exactly as the wires of the electric telegraph do.

513. The nerve-tubules are united together
into little ropes, or *"funiculi,"* and these
funiculi are again combined to make the
larger nerves.

("*Funiculus*" is the Latin word for a " *little rope*.")
The smaller nerves are merely tubular sheaths filled
with tubules. The sheath which contains the bundle of
tubules is a delicate structure, and is called the nerve-
bark or "*neurilemma*" (476). As the smaller nerves
are "bundles of tubules" (*Funiculi*)—so the larger
nerves are bundles of funiculi. In the large nerves the
bundles of funiculi are inclosed together in a common
sheath. In *Figure* 57, the composition of one of the
larger nerves is shown. The separate bundles seen
within the outer sheath are funiculi. Each one of them
consists of a quantity of such tubules as those drawn in
Figure 56, inclosed in the "neurilemma" sheath.

514. The funiculi of nerves *interchange their tubules*.

When the sheath of a large nerve is opened, it is found that the funiculi lying within are connected together in places, by threads passing between them. These threads are really tubules coming out of one funiculus, to pass into another. Although all the nerve-tubules are distinct from each other, from end to end, they get differently sorted and arranged in different parts of their course. This is, in order that tubules that come from distinct places, may be brought together in a common termination. It is exactly as when, in central telegraph offices, numerous wires come in from one line of railroad, and are then sorted and re-arranged so that some of them are sent off again in one direction by a new line, and some in another; wires that come in from different directions being at the same time associated together, and sent off in common. A rough diagram will help to make this more clear. Let 1 and 2 (*Figure* 58) represent separate funiculi, consisting of distinct tubules, but lying together in a common nerve-sheath for a certain distance, and then being sent off, each funiculus in a different direction, as at *A* and *B*. If there were no interchange

Fig. 58.

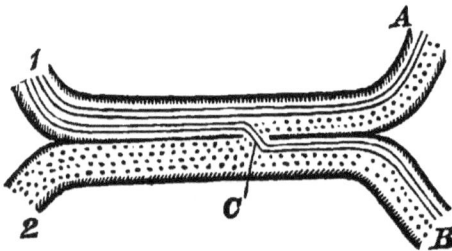

of tubules effected any where, between the funiculi 1 and 2, all the tubules from 1 would of necessity be distributed to *A*, and all those from 2 would be distributed to *B*. But if, on the other hand, an interchange of tubules is made at some intermediate point, as *C*, between the funiculi, then tubules coming from 1 are sent to both *A* and *B*, and tubules coming from 2 are distributed in the same divided way. The interlacing of the funiculi of nerves serves to bring different parts of the body into nervous communication with each other.

515. As numerous tubules are associated together in a common sheath to form *little cords* (funiculi), so all the various nerves are associated together in a common sheath to form a *large cord.*

All the numerous nerve-tubules, that are scattered to the hundreds of muscles of the body, really come from one large rope, or cord. The tubules of this primitive cord are first divided into large bundles, which are tied up together as nerves. Then, the nerve bundles are in a similar way divided into small bundles, which are tied up as funiculi. And lastly, each funiculus bundle is broken up into distinct tubules, which are scattered to their separate destinations.

516. The large cord, from which all the nerve threads that run to the muscles are given off, is called the *spinal cord.*

Fig. 59. It is so called, because it is contained in the interior of the back-bone or spine. The back-bone is hollowed out into one continuous cavity, from the head to the bottom of the back. In this cavity, the central cord of the nervous apparatus is packed out of harm's way.

517. There are *sixty-two* nerves given off from the spinal cord.

These nerves issue from the cord in pairs, one from each side, at successive portions of its length. The cord thus contains as many nerve tubules in its thickness, as the thirty-one pairs of nerves. *Figure* 59, represents the spinal cord giving off its row of nerves from one side.

518. The spinal nerves issue from the cavity of the backbone, *through*

small holes, left in the bony wall along each side.

The backbone is made up of a number of separate bony rings piled upon each other, and then connected together by fibrous ligaments (438). This is roughly represented in *Figure* 60, at *B*. The form of one of the separate bony pieces is shown at *V*. *O* is the interior of the ring, in which the spinal cords lies, when in its natural situation. When the several bony pieces are piled upon each other, the interior cavities are all continuous, so that one long chanel is formed. It will be observed that a notch is grooved into the rim of the bone at each side, (at *n* and *n*). There are corresponding notches in the under part of the rim of the bone which rests upon this, so that when the two pieces are in contact, small holes are left which pierce quite into the interior cavity of the canal. It is through these holes, that the successive pairs of spinal nerves pass out from the cord, each pair to be distributed to the parts of the body nearest to the points from which it startes. When the several pieces of the back bone are connected together in their natural positions, projecting points, corresponding with that which is marked *s* in the figure, lie along, one above the other, like a line of spines or thorns, which may be felt directly beneath the skin. The term spine is conferred upon the back-bone from this peculiarity; (it is derived from the word "*spina*," which is the Latin for a "thorn)." The back-bone is made up of several distinct pieces, in order that it may be capable of bending to a slight extent, instead of being an altogether rigid and inflexible column, which would have proved

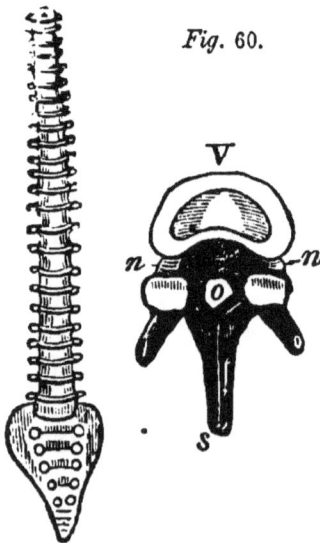

Fig. 60.

very inconvenient for many reasons. A small movement is allowed between each contiguous pair of bony
pieces. Each separate bone of the back is called a" *ver-
tebra*" from this characteristic, (the word being derived
from "*verto*," the Latin term for "to turn"). The back-
bone that is made up of these vertebræ, piled on each
other, is sometimes spoken of as the "*vertebral column.*"

519. The spinal cord is carefully covered
up in its bony canal by *coats of its own.*
The coverings of the cord are made of layers of serous
(453) and fibrous membrane (437), and serve to protect
the delicate bundle of nerve substance from any chance
mischief that might be done to it by the bending of the
back-bone. They are prolonged through the small side
openings, by which the nerves issue, in a tubular form,
and thus become continuous with the sheaths of the
nerves (513).

520. The spinal cord is composed of *two
different kinds of substance*, one white like
the nerves, and the other of a grey color.
The grey substance forms the centre
of the cord, and is entirely surrounded
by the white. When a portion of
the cord is cut across, its external
white and internal grey substance
are seen to be distributed somewhat
after the fashion shown in *Figure*
61.

Fig. 61.

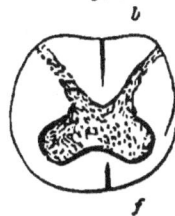

521. The outer white substance of the cord
consists of *nerve tubules.*
This white substance may be traced passing into the
several nerve bundles. It is strictly continuous with
them. The white portion of the cord is on this account
called its *fibrous* or stringy portion, to distinguish it
from the different material that it encloses. It properly
constitute the cord, viewed in the light of a collection
of threads.

522. The inner gray substance of the cord is principally composed of delicate *vesicles, or cells.*

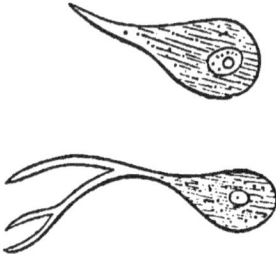
Fig. 62.

Hence it is called, for distinction's sake, the *vesicular* portion. *Figure* 62 shows the appearance of the cells of this portion of the spinal marrow, magnified 24,000 times. Most of them are of a lengthened pear-like shape, and end one way in a slender process or tail, more or less branched.

523. The cells of the vesicular part of the spinal cord are formed of very *delicate films*, and contain a *white pulpy substance* in their interiors.

The cells of nerve substance are of larger diameters than the nerve tubules. The smaller of them are about the same size as the blood corpuscles (339). Their films, and the matters which they contain, seem, however, to be exactly the same as the films and the white contained substance of the nerve tubules (508).

524. The films of nerve cells and tubules are principally delicate layers of *albuminous* material.

Nearly two ounces in every pound of nerve substance are albumen (242). This is the material that is employed in the construction of the walls of tubules and cells.

525. The white substance, contained within the nerve cells and tubules, is a peculiar combination of *fat, phosphorus, and water.*

In every pound of nerve substance, more than eleven ounces is water. In the same quantity there is nearly one-third of an ounce of phosphorus, and one ounce of fat. This water, phosphorus, and fat, are mingled together to form a sort of pulp, which is then deposited in the interiors of the cells and tubules. The fabrication of this pulp is of course the work of the living nerve cells. They separate from the blood the substances they require for its construction, and then mould them into this peculiar form as they pass them through their walls.

526. The cells of the vesicular portion of the spinal cord are placed *in the midst of meshes of capillary blood-vessels.*

A great number of blood-vessels enter the spinal cord; these are distributed, as a very finely meshed net of capillaries, everywhere through its vesicular substance. In the interstices of this net-work, the delicate cells are packed.

527. The *gray color* of the vesicular portion of the spinal cord (520), is due to the presence of the capillary blood-vessels.

The meshes of the capillaries are so fine, and the vessels themselves so numerous, that the white nerve substance is everywhere tinged by the color of the blood they contain. The delicate red streaks are invisible to the naked eye; but their color, softened by the white substance in which they lie, is perceived as a tinge of gray.

528. The gray vesicular portion of the spinal cord is very abundantly supplied with blood-vessels, because it has *active, vital work* to perform.

The nerve cells are more abundantly supplied with blood, than any other structures in the body. This is because their vital activities entail a more rapid waste and destruction of substance than any of the other textures suffer. As the waste of nerve substance is unusually rapid, an unusually abundant provision must be made for its repair.

529. The cells of the vesicular portion of

J

the spinal cord *originate the influence* which provokes the muscles to contract.

Wherever the nerves, that run from the spinal cord to any particular set of muscles, are divided, these muscles cease to be able any longer to óbey the directions of the will. The nerve tubules do nothing else but convey onwards an influence that is communicated to them. They have no originating power. Whenever force has to be originated, nerve cells and their abundant networks of capillary blood-vessels are provided.

530. The ends of the nerve tubules, and the cells of the vesicular part of the spinal cord are placed *in close communication* with each other.

The ends of the conveying tubules are brought near to the originating cells, in order that they may readily receive from them the influence they have to convey. Several of the tubules of the white, fibrous portion of the cord pass inwards and mingle with the substance of the vesicular portion. In some places, cells have been seen opening directly into the extremities of tubules. In *Figure* 63, a nerve cell is shown, leading at once into a nerve tubule in this way. In other cases nerve tubules have been seen to be simply looped about in close contact with nerve cells, without opening directly into them. Then the influence that the tubules receive must pass through both their own film walls and those of the cells.

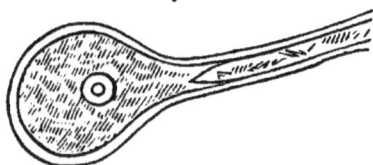

Fig. 63.

531. The nerve tubules are *formed out of simple cells.*

Simple, elongated, delicate cells are laid end to end, and then opened out into each other to form the tubules. Cells that are thus altered in form, and *developed* into tubes (429, 466), lose their principal vital powers, and cease to be able to reproduce themselves, or to accomplish any of the higher functions of cell vitality. Hence it is, that nerve tubules can only transmit influences that are originated elsewhere, and communicated to them. Nerve tubules, like capillary blood-vessels, perform scarcely more than mechanical functions. The one allows an influence, as the other allows a liquid, to flow along their channels. In the nerve tubules, however, the pulpy substance does not really move forward; it is merely some vibration, or subtile change of condition that is propagated onwards through it, as a wave is over the surface of water. The branching processes represented in the sketch of the nerve cells in *Figure* 62, are probably the commencements of tubules in course of formation.

532. Each nerve that is sent out from the spinal cord, possesses a *double root.*

One portion of the tubules, that enter into the composition of the nerve, is derived from the back of the spinal cord, and another portion is derived from the front of the cord. This is represented in *Figure* 64. *C* is a small portion of the cord, with its side turned towards the eye, *f* being its front, and *b* its back; *n* is one of the spinal nerves, coming out from the end by two separate roots, or bundles, 1 and 2; 1 being derived from the fibres of the front of the cord and 2 from those of the back. The two roots are united into one bundle, within the spinal canal, and then issue through the small hole left for their passage, as a single nerve.

Fig. 64.

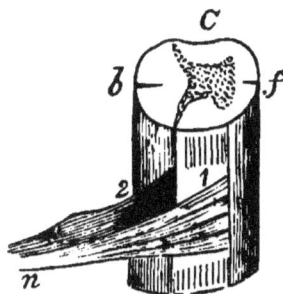

533. The back roots of the spinal nerves

have nothing to do with the conveyance of *motor stimulus to the muscles*.

When the front root of a spinal nerve is divided, certain muscles, supplied from it with nerve tubules, lose the power of contracting, under the direction of the will; but, when the back root of the same nerve is divided, no muscles lose their moving power.

534. The back roots of the spinal cord *convey sensations inwards*.

When the back root of a spinal nerve is divided, some *tract of skin*, supplied with tubules from it, loses its feeling. If the tubules of the back roots of the spinal nerves were traced onwards through the nerves to their ultimate terminations, it would be found that they ended principally in the skin, instead of in the muscles.

535. The back roots are composed of *sensitive, or afferent tubules*.

As these roots transmit impressions inwards, instead of outwards, the tubules and nerves that combine to form them are called in-carrying, or *afferent*. (The word "afferent" is derived from the two Latin terms, "*ad*"— "to" and "*fero*"—"to carry." The word "sensitive" is derived from the Latin "*sentio*," to feel). The various parts of the body feel by means of sentient tubules, which they send in to the spinal cord.

536. The sentient nerves originate principally *in the outer surface* of the body.

It is by means of sentient nerves that living animals acquire information concerning the external objects and relations that surround them; consequently, the extremities of these nerves must be distributed on the outer surface of the body, where external impressions can be made.

537. The skin is filled with the extremities *of sensitive nerves*.

The ends of the sensitive nerves are looped about in the skin very much in the same way as the terminations of the motor nerves are looped about amidst. the muscular fibres. *Fig.* 65 represents the arrangement of the terminal loops of the sensitive nerves in a portion of the skin of the thumb.

Fig. 65.

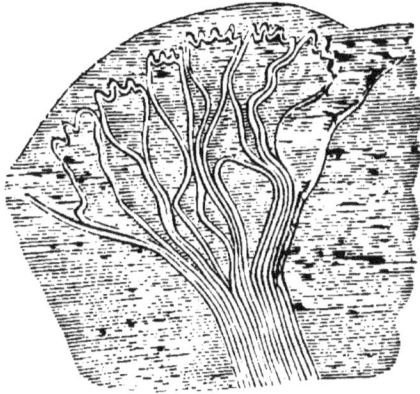

538. Most of the nerves of the body, therefore, consist of an admixture of *two different kinds* of tubules.

Motor tubules, that are distributed to the muscles, are gathered up together with sensitive tubules coming in from the skin. Both kinds are then bound up in one common sheath, and run on together, as a *nerve*, towards the spinal cord. But so soon as this nerve has entered the small holes in the side of the back-bone, its motor and sensory tubules are sorted out and separated from each other, and the former are sent forwards to one part of the fibrous portion of the cord, whilst the latter are sent backwards to another portion. "Afferent" tubules are also gathered, in smaller numbers, from the various internal organs that are generally devoid of sensation, but are capable of feeling pain when in a disordered state. Some come, too, from the muscles themselves, whose office it is to convey information of the condition of the various portions of the contractile apparatus, and of the manner in which the directions of the will are carried out. The motor tubules are called "*efferent*," in contradistinction to the *afferent*. (The word "*efferent*" is derived from the two Latin terms, *ex*, "from," and *fero*, "to bear.") The "efferent tubules" *carry influences out from* the spinal cord to the muscles.

539. The nervous apparatus thus enables the animal to *acquire information* concerning the external world, and to *modify its relations* with the same.

The impressions, made upon some portion of the external surface, are propagated inwards by the afferent nerve tubules, until they are registered upon the spinal cord and the inner shrines of sentient life, and then, influences are set at work in consequence of the information so received, which are sent outwards, along the efferent tubules to this, or the other set of muscles, so causing the mechanical apparatus of the frame to work and alter its position. The nervous apparatus is thus the most highly endowed, and the most important portion of the organization. It is that to which all the other structures of the body are subordinate and subservient.

540. Certain of the afferent nerves are fitted for the transmission inwards of *special and peculiar* sensations.

All the external portions of the body, and, in an inferior degree, most of its internal organs, are capable of feeling the contact and pressure of resisting bodies, and also the varying intensities of heat and cold. There are parts, however, that are organized and arranged in a very ingenious way, so as to enable them to catch impressions of a more subtle kind than those of mere touch and temperature. Thus the tongue is so organized that it can distinguish flavors; the nose, so that it can discriminate odors; the ear, so that it can discern sound; and the eye, so that it can mark the intensities with which different bodies send forth light. All these several organs of special sense, as they are called, have distinct nerves provided for their service, which are sent into particular parts of the centres of perception. These organs of special sense, however, are of so much importance as to require a separate description. (Chap. xx.). All that needs to be understood, for the present, is that they form a portion of the general plan, by which infor-

mation concerning the condition of external things is
acquired, and that their nerves belong to the "afferent"
set, adapted to carry external impressions in an inward
direction.

541. All the outer extremities of the sentient nerves are associated with *nerve cells
and capillary blood-vessels.*

The sensations are really all connected with and dependent upon the *origination of an influence.* When
the hand is pressed upon a hard, or cold body, the mere
contact or pressure is not a sensation. The contact provokes some change of condition in the nerve substance,
which change is then transmitted along the nerve-tubule
until it reaches a position within where it can be perceived. Now this change of condition can only *originate*
when there are nerve cells present to set up the action
(529). The nerve-tubules, it will be remembered, can
only transmit; they cannot originate. Layers of vesicular nerve substance (522) have been found to be present in most of the places in which external impressions
are communicated to sentient nerves, and there is very
good ground for concluding that they are also present in
other like positions in which hitherto they have not been
noticed. In this way, then, it appears that sensations are
impressions originated in vesicular nerve substance,
placed at the outer terminations of one set of nerve-
tubules, and transmitted along those tubules in an inward direction, and that motor force is an influence
originated in vesicular nerve substance, placed in contact
with the inner terminations of another set of nerve-
tubules, and transmitted along those tubules in an outward direction to the muscles. Such is, in rough outline,
the nature of the "nervous apparatus" of the frame.

542. Sentient impressions are vital changes
connected with the *destruction of nerve substance.*

A hard body cannot be felt, a sound cannot be heard,
a flower cannot be smelt, without some nerve cells being

destroyed in the finger, the ear, or the nose. Nerve substance is as much destroyed by the performance of its offices, as muscular fibres are by the performance of their contractile work (483).

543. All the organs of sensation are abundantly *supplied with blood.*

In the skin, the loops of the sentient nerves (537) are intimately mingled with meshes of capillary vessels. The same is the case with all other portions of the organization, where external impressions are communicated to nerve tubules. The vesicular substance associated with their extremities is *gray* with the presence of blood (527).

544. The blood furnished to the vesicular nerve substance carries to it oxygen, *to effect the corrosive decomposition* on which the nerve force depends, and nourishment to *repair the waste.*

The blood thus performs the same office for the nerve substance, that it does for the muscular substance (503.) A continued supply of blood is absolutely essential to the performance of the nerve functions. When the heart ceases, for a short time, to act, during a fainting fit, no blood is sent to the nerve substance, and all sensation disappears, as well as all motor power, until the blood again begins to be forced through the capillary vessels. The sensation of the skin is destroyed by cold, because the motion of the blood through its capillaries is arrested by this influence. A small quantity of blood is kept circulating amidst even the bundles of the nerve tubules to effect their renovation. Small blood-vessels enter the sheath of the funiculi of the nerves (513).

545. Nerve waste is *peculiarly rapid* during the employment of nerve substance.

Nervous substance is destroyed by its own efforts even more rapidly than muscular substance. It is of a far more ·delicate and combustible nature. Both the phosphorus and the fat, of which the contents of the nerve cells are principally composed (525), have very strong tendencies to unite with oxygen. In the production of nerve force, nerve substance is rapidly burned. It is in consequence of the higher combustibility, and the more rapid destruction of nerve substance, that all organs containing its vesicular matter are more abundantly supplied with blood than other structures of the frame. The repair must be as prompt and as free as are the destruction and the waste.

CHAPTER XVIII.

THE BRAIN.

546. The spinal cord is composed of *a gray mass of nerve cells*, inclosed within white fibrous substance.

The nerve cells are placed in the central portions of the cord, and are surrounded by white fibrous substance, which envelopes them in the form of a thick covering, or coat (520). The white fibrous portion is thicker than the gray vesicular portion inclosed within.

547. The gray vesicular substance of the spinal cord *is continuous* from one end to the other.

The vesicular substance of the cord is about 18 inches long. It forms one continuous tract within the white substance of the cord, but is thicker in some places than in others. It is thickest where nerves are given off that run to the very active parts of the frame. Where most

15   J*

nervous energy requires to be called into play, most
formative material is arranged. In some of the most
lowly forms of animated creatures, the nervous apparatus, that corresponds with the spinal cord, is formed of
separate masses of vesicular substance placed at wide
intervals, but connected with each other by a double
nerve thread.

548. The white fibrous substance of the
cord is *composed of nerve tubules* that are
merely extensions of those that compose the
funiculi of the nerves (513).

The outer white substance of the cord is merely a
collection of the extremities of nearly all the nerves of
the body brought together to be placed in communication with the vesicular substance. It is this collection
of threads, thus bound up together, that makes the
organ deserve its name of cord. The vesicular part
would cause the term *spinal marrow*, by which it is
sometimes known, to seem more appropriate. The vesicular portion of the spinal marrow does not run more
than two-thirds the length of the spinal canal; to the
remaining third the fibrous part alone extends, dividing
and subdividing into its nerve branches as it goes. The
spinal cord does not entirely fill up the cavity of the
bony case in which it is placed. It is covered with a
fibrous coat, (with a double serous one (454) intervening,
to allow of slight movements,) and is suspended carefully
and loosely by means of ligamentous ropes.

. 549. The spinal cord *receives impressions*
from without, and *emits motor force.*

The nerves, it will be remembered, are composed of
afferent and *efferent* tubules (538). The extremities of
both are brought into communication with the vesicular
substance of the cord. This vesicular substance receives
the impressions transmitted inwards by one set of tubules,
and originates the stimulant influence that the other set
transmits outwards.

550. The back part of the cord is concerned with the *reception of impressions*, and the front part with the *issuing of influences*. The afferent roots of the nerves have been seen to come to the back part of the cord, and the efferent roots to issue from the front of the cord (532, 533). Some of the tubules of the back roots are immediately placed in communication with the vesicular substance, which advances through the white external part to meet them. It may be seen doing so above the end of root 2 in *Figure* 64. Some of the tubules of the front roots are placed in communication, in the same way, with the vesicular matter, portions of which may be seen to advance to meet them, although less decidedly than is the case with the back roots.

551. The receiving bundles of the cord, and the issuing bundles, are divided from each other as distinct *back and front columns*. The white fibrous portion of the cord is divided into a right and left half, by two little cracks or fissures that dip into its substance. Each half is of course made up by the row of nerves that come from its own side (518). But each half is also further subdivided into three distinct columns (a back, a middle, and a front one), by the projecting portions of the gray vesicular substance that go to meet the roots of the nerves of each side (550). In *Figure* 61 all this is represented. The white substance of the cord is seen to be divided into a right and left half, and each half to be further divided by the arrangement of the gray substance into front, middle, and back bundles, or columns, marked respectively—*f*, *m*, and *b*.

552. The tubules of the fibrous portion of the cord are not *all made to communicate with the central vesicular part*.

Only a portion of the tubules, that pass in or out through the roots, are in communication with the vesicular projections that come to meet them. A greater number of them are bent upwards to form the front, middle, and back columns of the white part (551). These do not really belong to the spinal cord. They have nothing whatever to do with its vesicular mass, or active and originating substance. They merely pass along outside of it in close contiguity, bent upon a further journey. It will presently be seen where it is they go (561).

553. **The vesicular substance of the cord issues motor influence, when *provoked so to do by the reception of impressions* from without.**

The vesicular portion of the cord needs to be *provoked* to the performance of its peculiar functions, as much as the fibres of the muscles need to be *provoked* to the accomplishment of their contractile work (504). The stimulus, which irritates the vesicular matter of the cord until it issues motor force, is the reception of external impressions through the afferent nerves. When the spinal cord acts, it does so in answer to a suggestion made from without.

554. **The actions effected by the spiral cord are alike *unconscious and involuntary.***

When the spinal cord is provoked by impressions conveyed into its vesicular portion by afferent nerves coming from without, it produces movements that are devoid of sensation or consciousness on the part of the creature, and that are independent of all exercise of the will. Thus animals breathe, when they are insensible in sleep, without knowing anything about what is going on. Seven thousand times a large series of muscular cords contract powerfully and dilate again, between the time of a man's going to sleep at night and awaking in the morning. They do this in order that fresh air may be continuously pumped into the chest, throughout the night. It is the spinal cord that keeps up the incessant pumping. The presence of impure air in the air spaces of the lungs irritates the extremities of a set of afferent nerves that commence there. These nerves convey the impression to the vesicular portion of the cord, and this then issues the mandate, by a set of efferent nerves that run to the muscles of the chest, which causes them to come into active play. In all this, however, there is no trace of sense or voluntary effort.

555. The movements carried on by the spinal cord are unconscious and involuntary, because they need to be performed whilst *sensation and will are suspended* by sleep.

If breathing had been dependent upon either sensation or the exercise of will, it would have been suspended during sleep, when both of these powers are dormant. Other portions of the nervous centres, yet to be described, are refreshed by periodic intervals of repose; but the spinal cord never naps. It is always ready at its post to keep up such essential operations of life as cannot be left without care for an instant. When the spinal cord is injured, either by accident or disease, breathing stops, and the creature dies of suffocation within a few minutes.

556. The spinal cord takes charge of various operations in the body that would be less perfectly performed if *left to the exertion of the will*.

The spinal cord exercises supreme control over all the external openings of the body, and rigidly closes them against injurious influences. When an insect approaches the eye, the spinal cord makes its lid snap down with a wink. When any hard object, that has no business there, tries to get into the wind-pipe, the spinal cord ejects it summarily with a cough. If the intrusive body makes its lodgment in the nostril, the spinal cord dismisses it with a sneeze. When bland, nutritious sub-

stances are placed in the swallow, the spinal cord causes
them to be squeezed downwards into the stomach, and
then to be carried onwards in the digestive canal as the
process of digestion advances; or, if the material swallowed be such as the digestive organs are from any
cause unable to dispose of, the spinal cord reverses the
muscular movements, and the offending matters are
vomited back.   There are lowly animals that possess
no other vesicular nervous matter than that which corresponds with the spinal cord of more highly developed
creatures.   These animals swallow their food when it
touches their mouths, and withdraw their bodies from
influences that are injurious to them, without presenting
any indications of sentient life.   The spinal cord rules
over the department of unconscious operations.

557. The spinal cord *is prolonged* from its
own especial bony canal quite *into the cavity
of the skull.*

A large hollow case of bone is placed upon the top of
the back-bone, in such a way that the cavity of the spinal
canal is continued through a hole in the bottom of the
case, into its interior.   The bony case, placed on the top
of the spinal column, is called the *skull*, (a word derived
from " *skiola*," the old Teuton term for the head).   The
skull is, indeed, the solid frame-work of the head.   *Figure* 66 shows the appearance of the skull when sliced
through, in order that its cavity, *C*, may be exposed to

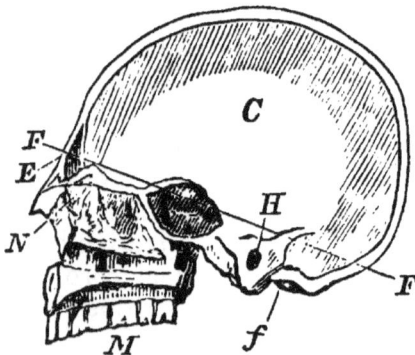

Fig. 66.

view.   *f* is the hole in
its floor, by means of
which the spinal canal
communicates   with
its cavity.   Through
this hole the spinal
cord passes upwards,
so that its upper ex-
tremity   lies   quite
within the cavity of
the skull.   This por-
tion of the spinal cord

is called the " prolongation of the spinal marrow." (Medulla oblongata.) It is thicker than the other portion,
and contains a greater abundance of vesicular substance.
This enlarged part of the spinal cord it is, that takes the
important movements of breathing and swallowing under
its particular charge.

558. Above the upper termination of the spinal cord is placed
*another mass of vesicular nerve
substance* inclosed within white
matter.

In *Figure* 67, *M* represents the enlarged termination of the spinal cord
(557), and *S* another mass of vesicular
nerve substance situated still further in
the cavity of the skull. This mass is really subdivided into subordinate portions,
but may be spoken of as one substance,
on account of these being all associated
together in the performance of similar
offices.

Fig. 67.

559. The vesicular mass placed above the
termination of the spinal cord is concerned
with the performance of actions that are
attended by sensation.

All the nerves that bring in impressions from the
organs of special sense, the eye, the ear, the nose, and
the tongue (540), run to this mass. This at once points
out what its office must be.

560. The vesicular substance, which is concerned with sentient operations, and which
receives the nerves of special sense, is called
the sensorium.

The different portions of the mass are in connection
with the separate nerves of sensation, and are marked
by distinct names ; but it is convenient to speak of the

whole as forming one common centre of sensation and
sensory acts. (The word "*sensorium*" is derived from
the Latin "*sentio*," *to feel*, and means properly the seat
of feeling. In *Figure* 67, *M* represents the enlarged
termination of the spinal cord, and *S* is the sensorium.

561. Some of the *nerve tubules of the spinal
cord* run up to connect themselves with the
vesicular mass of the sensorium.

This, then, is what becomes of a part of the nerve tubules that have been before seen not to end in the vesicular substance of the cord (552). The fibrous columns
of the cord pass up outside its inner vesicular substance,
until they reach the sensorium. It is only the vesicular
portion of the cord (its vital, or originating part.) that
really terminates at the point marked *M* in *Figure* 67.
By its fibrous part, the cord is continuous with the sensorium, placed above, as represented in the figure. Some
of the tubules of the fibrous columns are brought into
communication with the vesicular matter of the sensorium, that they may transmit impressions to it, and receive influences from it. Hence, the sensorium is, like
the spinal cord, composed of a central gray vesicular
portion, and of an external white tubular portion.

562. The vesicular masses of the sensorium
are placed in the head, in order that they
may be *near to the organs of special sense.*

The head consists of *the skull,* which contains the vesicular nerve matter of the sensorium, (and other still
more important organs yet to be noticed, 569,) and of
the face, which is merely a bony frame-work, where the
extremities of the nerves of special sense are conveniently
distributed for the work they have to perform. This is
shown in *Figure* 66, where *C* is the cavity of the skull,
and all below the line *F F* is the bony frame of the face,
composed of *E*, the hollow in which the eye, or organ of
vision is placed; *N*, that of the organ of smelling; *M*,
the mouth, with its organ of taste; and *H*, the organ of

hearing. The sensorium is the additional piece of vesicular nerve matter which is necessarily furnished for the perception of sensations and the management of the movements connected with them. Such animals as are nearly devoid of organs of special sensation are also destitute of a sensorium. The two are inseparably connected, and as the sensorium when present is placed in the head, the nerves of special sense that minister to it are placed in front of the same part of the body to be near at hand. In the lower animals that are without internal skeletons of bone, and therefore that have no skull, the sensorium is still situated in the head, immediately behind the organs of special sense.

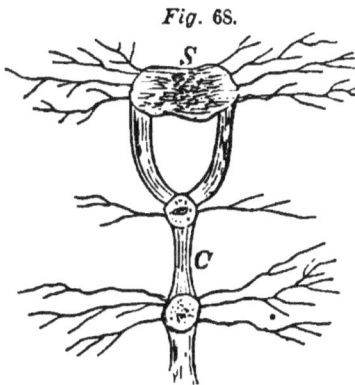

Fig. 68.

Figure 68 represents the sensorium of the insect. C is the last portion of the vesicular nerve substance that takes the place of the spinal cord, and that rules over the unconscious acts. Two bundles of nerves proceed forwards from this to a large mass of vesicular substance, S, placed in the head of the insect. This is the sensorium, and it distributes nerve branches to the antennæ,—which are at once the ears and fingers of the creature,—to the mouth, and to the eyes. The head thus is the seat of sense, as the abdomen is the seat of the digestive operations (299).

563. The movements effected under the influence of the sensorium, are *instinctive* and in no way connected with the exercise of the will.

(The word "*instinct*" is derived from the Latin "*instinguo*," to "*incite*" or "*provoke*.") The instinctive actions are such as are provoked or called forth by the impressions transmitted from the organs of sense. The

sensorium causes certain movements to be made in immediate reply to sensory impressions communicated to
it, and these movements as directly and necessarily follow the stimulus of the sensation as the unconscious
acts follow the stimulus of the vesicular masses of the
spinal cord.   In both cases the movements are strictly
provoked movements, and are called forth by the application of external influences, and not by the will of the
creature.   But, in the one case, the movements are unconscious; the creature knows nothing whatever about
them.   Whereas in the other case they are conscious,
the creature being fully aware of them, although quite
unable to alter or modify them.   The instinctive actions
are involuntary on the part of the creature, but are
attended either by pleasure or pain.

564. Creatures that have no higher development of nervous organization than the sensorium, *are governed by instinct alone.*

This is the case with insects, and with all animals that
are lower than them in the scale of nature.   The insect's
nervous apparatus consists exclusively of conveying
nerves, of organs of sense, and of a chain of small vesicular masses that run the entire length of the body, and
are connected together by bundles of tubules passing
between them.   Of these vesicular masses, all excepting
the foremost one, which is placed in the head, are concerned in keeping up movements essential to the continuance of life, but of which the creature is unconscious.
That foremost one, however, confers upon the creature
conscious life, and makes it a pleasure for it to do some
things, and a pain for it to do others, and so gets to
modify and control the actions of the rest.   The insect,
in consequence of possessing a sensorium, but no other
organ of yet higher power, is driven by pleasure and
pain to the performance of certain operations which
make up the sum of its life.

565. Instinct is *unvarying* in its operations.

In this way it is at once distinguished from intelligence

and reason. It has no perception of an object in view, and no power of modifying its actions to meet variation of circumstance. So far as the operations of instinct are concerned, the perception of the object to be gained is in the mind of the great First Cause who plans the organization, and adapts the external influences of nature to the organs so as to bring out through them certain sure results, rather than in the creature. The creature is strictly *an instrument*, being constantly played upon by a very skilful hand, so that wonderful harmonics are called out from it, independently of all knowledge of its own. But every instrument that is made in the same way does the same thing. Every bee makes its waxen cell exactly like its neighbor, and gathers and stores its honey after the same fashion. In creatures that are exclusively governed by instinct there is no difference of character or temper, as in more reasoning beings. Every bee is alike clever, industrious, and thrifty. Every wasp is alike bold and fierce. Insects afford the best illustrations of the nature of instinct, because they are the most highly developed of all the animal races that are destitute of reason.

566. Instinct cannot be *trained or educated*.

The insect springs into existence, perfect in all its work. The bee flies off the instant it emerges from its nursery, and lights upon the first flower that offers; collects its store of sweets, and flies back to the hive to deposit it in its waxen cell. The bee of five minutes old transacts all the ingenious operations of its craft as skilfully and as well as the bee of six months old. And no bee can be trained to do anything which it does not perform in the first instants of its life. In instinct, experience goes for nothing, and education is altogether impossible.

567. Man is *influenced by instinct* as well as the lower creatures.

The masses of vesicular nerve substance that are placed above the upper extremity of the human spinal

cord, produce certain influences on the actions of men.
The passions and emotions are all instinctive impulses
called forth by impressions of sense. The wasp attacks
any one who attempts to meddle with it, under the impulse of true passion. When men act under the influence of rage or anger, they do, to a great extent, the
same thing as the wasp.

568. The instinctive impulses of men are
modified and controlled by the exertion of
higher faculties.

The emotions and passions of mankind are more or
less governed by reason, and the sense of right and
wrong. Man is designed to be a rational creature,
although there are instinctive propensities mingled with
his nature.

569. The instinctive impulses are controlled
in man through the instrumentality of a *mass
of vesicular nerve substance, which is placed
within the skull above the sensorium.*

Fig. 69.

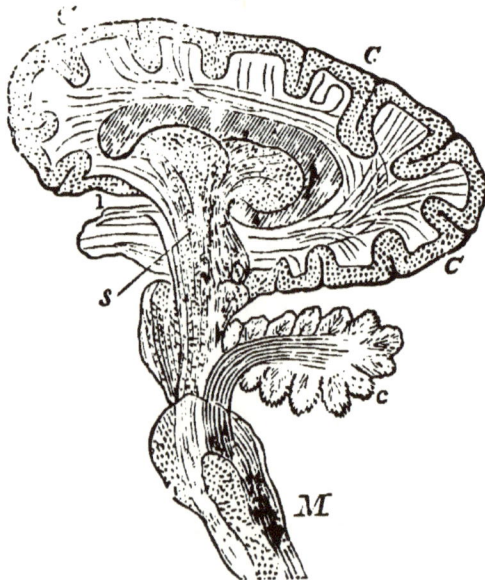

The sensorium is completely hidden in man, by a large mass of nervous substance which is covered over it. This is represented in *Figure* 69, where *M* is the upward prolongation of the *spinal cord*, *S* is the sensorium, and *C*, *C*, *C* is the larger and higher mass, which envelopes the sensorium within its substance.

570. **The nerve mass, placed above the sensorium of man, is called the *cerebrum*.**

In *Figure* 69, *C* is the "*cerebrum*" (the word is the Latin term for "*brain*"). In common language, the connected mass of nerve substance, which is placed in the cavity of the skull (*C* of *Figure* 66), is called the brain. But, properly, this brain is made of three distinct organs, the *vesicular substance* of each of which is separated from that of its neighbors, although the whole are united together by white tubular fibres. The three several parts of the brain are the top of the spinal cord (*M*) [*Figure* 69], the sensorium (*S*), and the cerebrum or *brain proper* (*C*). It should also, however, be remarked that there is a fourth organ of a subordinate nature comprised within the brain. This is marked *c* in *Figure* 69. It is called the "little brain" or cerebellum. The cerebellum sends down a large bundle of nerve tubules, which help to form the back columns of the fibrous portion of the spinal cord (533, 551). These, it will be remembered, are continuations of the back or sentient roots of the spinal nerves (550). The precise office the cerebellum has to perform is not yet deemed to have been fully determined; but it is very probable that it receives impressions from the several parts of the muscular apparatus, and so keeps a sort of register of the condition in which the moving organs are at any given instant. Without such information being present to the mind, it would be quite impossible that the mechanism of the body should be made instrumental in carrying out the purposes of the will. The size of the cerebellum is generally proportioned to the moving power of the creature.

571. **The cerebrum is the heart of the brain**

which is concerned in all the operations of *intelligence, reason, and moral purpose.*

By means of the super addition of the cerebrum to the sensorium, the conscious creature is made also intelligent, rational, and cognizant of moral obligations. As the spinal cord is the seat of unconscious operations, and the sensorium of sense and conscious actions, the cerebrum is *the seat of mind.*

572. All creatures, that possess *any trace of cerebrum* in their nervous organization, are capable of performing *some kind of intellectual operation.*

The insects have no trace of cerebrum in their organization; hence, they are altogether devoid of intelligence, and entirely impulsive in their actions. Dogs and horses, on the other hand, have cerebral masses in their brains; hence, they are possessed of a certain amount of intelligence.

573. The amount of an animal's intelligence is generally *in proportion to the size of its cerebrum,* as compared with the other vesicular portions of its nervous apparatus.

In *Figure* 68, a sketch is given at *S* of the insect's brain, which is entirely composed of sensorium, without the addition of any trace of cerebrum. In *Figure* 70, 1 represents the brain of the cod-fish, *S* being the sensorium, *c,* the cerebellum, and *C* the cerebrum, placed upon the sensorium. It will be observed that the sensorium and cerebellum make up together many times the mass of the cere-

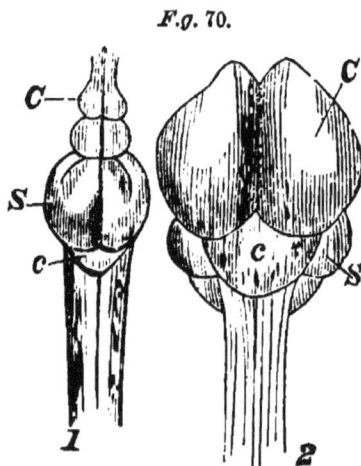

F.g. 70.

brum. 2 is a sketch of the brain of the buzzard. It will be noticed here that C, the cerebrum, is so large that it hides a great portion of S, the sensorium. It will also be observed how large in this brain the cerebellum, c, is. This is because the bird is one of great muscular development, and vigorous power of flight (570). In *Figure* 69, it will be observed that in man— the most intelligent of all the animal creation—the cerebrum is very much larger than the sensorium. Just to this extent he is designed to be more intelligent and rational, than instinctive. The cod-fish, on the other hand, is designed to be more instinctive than rational, to the extent to which its sensorium (S) is larger than its cerebrum (C).

574. The extent to which the *skull is larger than the face*, generally indicates the preponderance of intelligence over instinct.

Creatures with large skulls and small faces are generally intelligent. Creatures with large faces and small skulls are as generally impulsive and irrational. If the skull of any animal be sliced through as represented in *Figure* 66, the size of the space above the line *FF*, as compared with the space beneath it, will express the degree of intelligence, as compared with the force of instinct. The reason for this is that the face is the frame on which the nerves of sensation are distributed, and these nerves are appendages of the sensorium, and are large or small in proportion as it is large or small. Hence, a large face and small skull indicate a large sensorium and small cerebrum ; or, in other words, strong instinct and weak intelligence.

575. The cerebrum, like the spinal cord, is composed of *gray vesicular* nerve matter and *white tubular* substance.

The gray vesicular substance of the cerebrum is alone concerned in originating influence. The white fibrous portion only transmits from place to place, what the gray vesicular portion has produced.

576. The gray vesicular substance of the cerebrum is *placed outside of the fibrous white substance.*

When the brain is cut through, it is found that the central portions are white, unlike those of the sensorium and spinal cord (520, 561), and that they are surrounded by a layer of gray substance. This is represented in *C* of *Figure* 69. The reason for this difference of arrangement of the vesicular and fibrous portions of the cerebrum, as compared with the other nervous organs, is that the cerebrum is the highest of the series. There is nothing above it, to which tubules need to ascend. The tubules therefore end in its mass, and the active cells of the vesicular substance are laid over their terminations. This also allows more space for the vesicular substance than could be afforded if it were packed away in the interior of the organ.

577. The vesicular substance of the human cerebrum has a *very wide extent.*

It has been estimated that, if the vesicular layer of the human cerebrum were spread out flat, it would present a surface containing no less than 670 square inches.

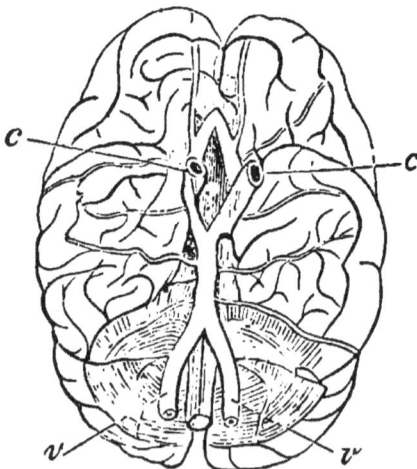

Fig. 71.

578. The vesicular layer of the human cerebrum is *folded up into convolutions,* for the convenience of package.

The appearance of the folds, or *convolutions* of the cerebrum is represented in *Figure* 71, sketched as seen on the under surface. The vesicular layer is crushed up together, in order that its wide expanse may be packed away into the cavity of the skull, much as a handkerchief is crushed up by the hand when about to be placed in the pocket. The lines in the figure are the inward dipping folds.

579. The vesicular portion of the cerebrum is the *seat of its power.*

It is the cells of the brain that form its active part. They are very much of the same nature as those already described (523). The greater the proportion of vesicular substance any cerebrum contains, the more active it is. This, in a general way, is marked by the number of convolutions the surface of the organ presents, because the more expansive the layer that has to be packed away in the cavity of the skull, the greater must be the number of folds into which it is arranged. There are a vast number of folds in the cerebrum of man, and comparatively few in that of the lower animals. In the lowest of all which possess a distinct brain, the cerebrum is smooth and quite devoid of convolutions.

580. In man, the brain forms a *thirty-sixth part of the entire body.*

And of this brain, by far the largest portion is the cerebrum, which is concerned in the support of mental operations. The cerebrum of man is four times larger than all the rest of the nervous substance taken together. This indicates how important an organ the brain is, in the human frame, and how much power mind is destined to possess over instinct. In birds the brain forms only the *two hundredth* part of the body; in reptiles the *thirteen hundredth* part; and in fishes the *fifty-sixth hundredth* part. The adult human brain varies in weight from two to four pounds. The brain of the idiot sometimes does not weigh more than twenty ounces.

581. The *large* brain of man is furnished with a VERY LARGE *proportion of blood.*

This again indicates how important the organ is in the

16   K

human frame. In proportion to its activity must be the
quantity of the blood that is sent to it, to furnish the
oxygen which sets up the destructive decomposition of
its substance, and to supply the material which has to
effect the repair of the waste (544). The substance of
the cerebrum is consumed by the efforts of thought, as
the substance of the muscular fibres is consumed by contractile movement. The brain is the *one thirty-sixth part*
of the entire body, but this one thirty-sixth part actually
receives to its own share *one-fifth part* of the entire
quantity of blood devoted to the nourishment of the
frame. The brain is, therefore, *seven times better supplied*
with material support than the other organs, and is seven
times more active than they are. The brain in fact is
the grand and supreme effort of the organization. It is
the part of the body, to whose service, in man at least,
all else is subservient. Every part of the human frame
has been made in order that it may become the residence
and the instrument of mind. As sensation and voluntary
movement is the highest privilege of animal life, mind
is the highest privilege of human life.

582. The blood, which supports the activity of the brain, is supplied by vessels that
branch abundantly over its outer surface.

As the vesicular part of the cerebrum, which is its
vitally active portion, is situated outside the white tubular part, the blood can be most conveniently furnished
for the support of its activity, by the spreading of arterial vessels externally to it. The brain in fact has a
great coat of blood-vessels, which branch over its external surface, and send fine meshes of capillaries down
between its folds and in amidst its cells, in all directions.
Figure 71 shows the distribution of the blood-vessels on
the outer and under surface of the brain; *c c* and *v v* are
the great arterial branches which come in to the bottom
of the brain through the floor of the skull. There are
four of them, and they unite to form a sort of circle,
from which myriads of other branches are then sent off.
On the outside of the blood-vessel coat of the brain,

there is a double coat of serous membrane (454) and a strong coat of fibrous membrane (437). The fibrous coat connects the organ firmly with the walls of the skull, and the double serous membrane is then interposed between this and the soft delicate substance of the brain, to allow a slight degree of motion to take place.

583. The white substance of the cerebrum consists of *bundles of conveying tubules.*

These tubules come up in the interior of the organ and run into the external layer of vesicular matter, some of them communicating with its cells, and others looping about amidst them (530). A portion of these tubules convey impressions to the vesicular substance, and a portion carry out influences that are originated there.

584. Some of the tubules of the cerebrum come from, and return to, the *vesicular masses of the sensorium.*

These bundles of tubules are shown at 1 in *Figure* 69. They serve to establish a communication between the sensorium and cerebrum ; but they are not fibres or tubules, continued up from the spinal cord and sensorium to form the white substance of the cerebrum. They are short distinct tubules, merely passing between the vesicular masses of the two organs. The cerebrum is, as it were, an altogether independent piece of organization added to the rest. It is not a portion of the nervous frame that is essential to animal life, but is something superadded to it, in order that animal life may be endowed with mind.

585. The tubules, that run from the sensorium to the cerebrum, carry sensations that are designed to be made the *subjects of mental operations.*

The mind works upon sensations in the first instance. It acquires *primarily* the elements, upon which it founds all its varied operations, through the instrumentality of the organs of sensation. When any of the external organs of sense are imperfect, the mind grows more slowly, and is ministered to less effectually. There is great probability that the mind would remain dormant and undeveloped through life, in any creature that was placed in the world deficient in every external capacity for receiving the impressions of sense.

586. The tubules, that run from the cerebrum to the sensorium, *convey the results of mental operations to the seat of consciousness.*

In this way man has, as it were, a double consciousness. He has the consciousness of a lower, and that of a higher life. The sensorium notices what is taking place without in the external world, and what is taking place within, in the mind. It does this through the tubules that come in from the organs of sense, and through those that come out from the cerebrum. (The word "*conscious*" is derived from the Latin term "*conscio*," to know.)

587. Some of the tubules of the cerebrum run *from one part* of its vesicular layer *to another*.

These bring the different portions of the vesicular substance into communication with each other. They carry force that originates in one part to other parts, to become operative there in exciting a new activity. They assume the appearance of bands of white matter crossing the interior of the cerebrum in various directions.

CHAPTER XIX.

THE OPERATIONS OF THE MIND.

588. Through the instrumentality of the cerebrum states of consciousness *are changed into ideas.*

Consciousness is simply and purely an internal state. It knows nothing whatever of the external world. So soon as any state of consciousness becomes connected with the notion of some external object that has had to do with the production of the state, it becomes an idea. This conversion of the conscious state into an idea, is, then, one of the offices which the vesicular substance of the cerebrum performs. The *conscious* sensorium transmits a certain force upwards to the cerebrum, and this force excites in it a new state of activity, which in turn becomes an *idea.* (This word is the Greek term for a "*vision*" or "*appearance.*" *An idea is a vision or image in the mind.*)

589. The *perception* of an idea is an *act of consciousness.*

The cerebrum, made active by the transmission of force from the sensorium, reacts upon it by sending down to it an influence, which awakens in it a further state of consciousness. This, however, is now *perceptive consciousness:* it is the consciousness of an idea, instead of a sensation. Thus, conscious states of the sensorium may be called up either from without or from within. A change in the nervous substance of the eye produces in the sensorium a consciousness of a sensation, and a change in the nervous substance of the cerebrum calls up in it a consciousness of an idea. ("*Perception*" is derived from the Latin word "*percipio,*" which means to "*take up,*" or "*seize*" entirely.)

590. A conscious state is changed into an idea *through the faculty of attention.*

Unless the attention is fixed upon a sensation, it does not become a perception. Persons, who are sleeping soundly, will sometimes start when suddenly exposed to the influence of a strong light, and yet will know nothing whatever concerning the cause of the disturbance. The light is seen, but not perceived. "*Attention*" is another of the offices which the cerebrum performs. (The word is derived from the Latin terms "*ad*" and "*tendo,*" which mean literally to "*stretch* or *bend to.*") Attention is a bending of the mind to the state of consciousness present. It is, so to speak, a throwing of the force of the cerebrum into the sensorium, in order that a conscious state may be changed into an idea.

591. Ideas, when once formed, are *fixed by the cerebrum as memories.*

Every sensation, or conscious *state* that is perceived, is recorded by the cerebrum, and kept in such a form that it can be recalled, or reproduced, at a future time. (The word "*memory*" is derived from the Latin term "*memor,*" which means "*keeping in mind.*") It is probable that no single state of consciousness is ever again altogether lost, *when it has once been fully and perfectly perceived.*

592. Ideas are capable of being *connected with simple states of consciousness,* either of the nature of pleasure or of pain.

The connection of an idea with the sense of enjoyment or discomfort, converts it into what is called *an emotion.* The desires, and the moral feelings and sentiments, are all of this compound nature. They are called *emotions* because they powerfully stir or *move* men to act in particular ways. (The word is derived from the Latin term "*emoveo*," which means to "*stir*" or "*remove*.") The emotions are really instinctive impulses colored and tinged by the influence of ideas (568). They are the actions of the sensorium, influenced by the operations of the cerebrum (569). It is not possible for any human creature to do aught that is purely instinctive, on account of the presence of the highly active cerebrum in his organization. As soon as a conscious state is produced in his sensorium, through the influence of an impression of sense, that conscious state calls up an idea in the cerebrum, at the same time that it tends to produce a *movement*. The movement that really follows is, therefore, a result of a combined influence, derived half from the sensorium and half from the cerebrum. It is, in the majority of cases, partially an instinct and partially an operation of mind. What, in the lower creature, is the sole guide of action, becomes, in the human creature, merely a *motive*, or prompter of activity, and is allowed to operate, or is prevented from doing so, according to circumstances. In man, the compound motions, like the simple sensations, operate upwards upon the cerebrum, as well as downwards upon the muscular apparatus. They *stir up* mental activity, as well as bodily; but, in a general way, their force is mainly expended in the one or in the other direction. The more people's emotions lead to demonstrative actions, the less they conduce to intellectual growth; and, on the other hand, the more thought they excite, and the more enduring result they lead to, the less they immediately show.

593. **The results, that are worked out through the activity of the cerebrum, are termed collectively "** *The mind.*"

The word "*mind*" is derived from an old Anglo-Saxon term "*munan*," to "*mark*" or "*note*." The mind is that which marks or notes. The Anglo-Saxon "*gemynd*" was originally adopted as a convenient generic name for all these mysterious powers of thought, which distinguish men from the rest of the animated creation.

594. Nothing is really known of the *true nature of mind.*

Mind continues to be *the great mystery of life.* Its operations are dependent upon the cerebrum ; for, whenever this organ is injured, the mind suffers in proportion. When the cerebrum is destroyed, the mind ceases to be able to make any further demonstrations of its existence. This, however, proves nothing more than that the cerebrum is the material instrument through which the mysterious agent manifests itself. It is in this sense that the cerebrum is appropriately said to be the seat of the mind.

595. Mind is distinguished from all the more material forms of vital force in *not consuming the food which ministers to its activity and strength.*

The muscles destroy the plastic constituents of the blood, and get their force out of them (483). The organs of sense and the sensorium do the same (542). So also does the cerebrum, in keeping up the connection between the mind and the nervous organization. It wastes abundance of vesicular nerve substance in converting conscious states into ideas, and in changing ideas into concious states (581). But ideas properly constitute the food of the mind. This food the mind does not destroy in getting its force out of it. On the contrary, it stores these ideas up as fast as it can get them, and then uses them over and over again, putting them together in ever new relations, and making ever new combinations out of them. The food of the mind is itself immaterial and indestructible. When once a mind has acquired a certain store of ideas, and has registered them as memories, it may be entirely shut out from the external world, by the loss of all its ministering senses, and still it will be able to go on growing and increasing in vigor and strength.

596. Mind possesses an *unlimited power of growth.*

No one has yet been able to conceive a limit to the power of the human mind. When properly exercised, and well employed, it seems to possess the faculty of going on growing so long as it remains in communication with the corporeal frame. There are cases on record, in which men 80 years old have still retained the mental vigor of their early life, and have been able to add day by day to their intellectual stores. This indefinite expansiveness of mental power immediately results from the facts stated above, that every new idea becomes registered as a memory, when once fully perceived, and that the mind works upon ideas without destroying them. With the mind it is not, as it is with the corporeal organs, a constant consumption of material to repair its own waste. It is a continued growth effected by the preservation and heaping up of every fresh acquisition that is made. No one can ever form the most remote notion of the end he may ultimately reach when he starts in the race of intellectual life. Therefore, no one ever gives his own capacity fair play who does not *keep his mind at work as long as he can.*

597. Mind is capable of *acting upon the nervous organization.*

In this sense, mind is a true force. It exerts a certain influence over the conditions of the sensorium and spinal cord, and therefore over the movements that are originated in these. Nervous force and mind force are intimately related, much in the same way as nervous force and muscular force. Either can excite the other. Nervous force acts on mind, when sensations call up ideas; mind force acts on nervous force, and through it on muscular force, when ideas produce perceptive consciousness and corporeal movement.

598. Mind force, in the state of exertion, is termed *the will.*

K*

(The word "*will*" is derived from the Anglo-Saxon "*willan*," itself probably taken from "*wol*," "*good*"— *that which is desired*.)

599. The will, therefore, is altogether an *independent power*.

It is force of will that makes man a free agent. It is by will that mind acts on the body, and it is, therefore, through it that the body is removed from the rule of the instincts, and placed under the direction of reason and intelligence. The faculty spoken of before under the designation of attention (590), is one of the forms in which will is exerted.

600. The will possesses the power of *controlling the operations of the mind*, as well as those of the body.

That is to say, the whole force of the mind is able to be concentrated upon any one of its operations. Man is able, not only to will to make certain movements, and to do certain things, he is also able to will to enter upon certain lines of thought, and to perform certain mental operations, upon which the development of intelligence depends.

601. All the mental operations consist in *associating ideas together in various ways*, and *in forming new complex ideas* out of these associations.

The mind operates entirely upon ideas, and builds up its results out of them. It places certain ideas, that have been separately acquired, together, and so fixes them in the memory that the present consciousness of one will, at any time, recall another into a state of consciousness also. The formation of ideas out of sensations is the work of the cerebrum. The grouping and arranging of these ideas, through the exertion of the faculties of attention and will, and the eliciting of yet other ideas out of their relations, is the work of the mind. In this way it is that the cerebrum stands, as it were, between mind and material organization. Without the cerebrum, there can be no ideas formed out of sensations, on which the mind can begin its work.

602. There are *two main principles on which the mind works* in associating ideas together.

That is to say, there are two ways in which it fixes ideas together in memory, so that the present consciousness of one may, of necessity, call back the consciousness of another.

603. The mind fixes together in the memory ideas that are *constantly presenting themselves in natural companionship.*

Thus the mind inseparably associates together the idea of greenness with the idea of grass; so that the notion of grass can never be recalled in the future, without the thought of its green color, of necessity, coming back with it too. This kind of association has been called the *association of connection, or of contiguity.* Out of it spring expectation, and belief in unvarying causation. The mind gets to expect that things, which it has always found to be connected together, will for ever continue to be so connected. From its experience it forms its anticipations of the future.

604. The mind fixes together in the memory ideas that *present some features of resemblance.*

By this means ideas that are subjects of present consciousness, are made to suggest or bring back other ideas that possess some kind of similarity to them. By this kind of association, like ideas are kept in the mind in groups. Hence, the plan is called the *association of similarity.*

605. *Recollection* acts by means of associations that have been previously formed.

By recollection, a consciousness is sought that may through one of the above means of association, serve bring back some other state of past consciousness that is desired to recall. (The word is derived from the Latin term "*recolligo*," which means to "*gather up again*.") By recollection, ideas that have been laid aside are gathered up, in the hope that some of them may suggest the notion that it is desired to recover.

606. By means of its associations of ideas, the mind *reasons out* certain general results.

(The word "*reason*" is derived from the Latin term "*reor*," to "*judge*.") As the will is the *mind acting*, the reason is the *mind judging*.

607. The general results, which the reason draws from the association of ideas, are called *facts or truths*.

(The word "*fact*" is derived from the Latin "*facio*," "*to do*.") It means literally a thing done, or reality. "*Truth*" is properly the third person of the verb "*to trow*"—i. e., "*troweth*." ("*To trow*" is derived from the Anglo-Saxon "*treowian*," "*to trust*.") A fact, or truth, is a notion that presents itself to the mind as a reality, and is trusted in. All intellectual effort has for its object the discovery and perception of truth or reality. (The word "*intellect*" is derived from the Latin term "*intelligo*," which means to "*understand*.")

608. The reason seeks truth by *comparing ideas together*, and then forming new notions out of the comparison.

The reason judges of two or more ideas, when placed in mutual relation, how far they are like each other, and how far they are unlike. Comparison is thus a more advanced operation than association. Association hangs one idea upon another, by force of habit, or by the perception of resemblance; but comparison determines how far ideas, that are hung together, really belong to each other, and to what extent one has characters that do not belong to the other.

609. The reason *identifies such ideas as are distinct.*

Having compared ideas together, the reason marks the distinctness of each. The word " *identify*" is derived from two Latin terms, " *idem*" and " *fio*" which signify " *the same*," and " *to be made*," or " *esteemed*.") A thing identified is one esteemed, or determined to be the same. Identification is a very important operation in reasoning. There can be no safe result arrived at, until every distinct idea has been recognized as such, and has been registered according to its exact worth. Confusion of ideas with each other is one of the most common grounds for men's failure in the pursuit of truth.

610. The reason *builds up* simple distinct ideas *into general notions.*

This building up of precise ideas into comprehensive general notions of truth, is the highest and most important work the intellect performs. The mind having acquired a large stock of ideas, and having made these ideas precise, is always at work sorting, rearranging, and reconsidering them in every possible variety of form, deducing or drawing out still newer and more complex ideas from the simpler ones. This building up of complex ideas is called *the process of generalization.*

611. The intellectual processes are carried on *almost involuntarily.*

The sole object of the operations of pure intellect is the ascertainment of truth. Hence, it would not be well that the will should have too much to do with the process, or it would be likely, considering the various influences by which it is affected, to lead as much *from* as *towards* the desired end. All the conclusions of intellect are involuntary so long as they are free from the operation of motives. The convictions must be exclusively deductions of the judgment, if they are to be worthy of trust.

612. The will *fixes the attention on the ideas* that are suited to lead the thoughts into the appropriate train of suggestions and associations.

When once the attention has been firmly fixed upon some idea that connects itself with the subject under investigation, the rest may be mainly left to the laws of association. The mind will then go on working of its own accord in the desired direction, building up its ideal structures, and eliciting from them its wonderful results. Studious men need to make an effort to sit down to their work; but, having once made this, little more is required *until the mind begins to be fatigued with its labors.*

613. The will *prevents the intrusion of ideas* that are foreign to the subject under consideration.

The power of employing the will in this way is termed the faculty of *abstraction.* This is a most important faculty to the pursuit of knowledge. Unless it is possessed, every chance sensation is apt to call up ideas that will associatively run away with the thoughts into new channels, and so divert them from the one which it is desired to pursue. The will, therefore, chiefly aids the intellect by setting its forces at work in the right direction, in the first place, and by then guarding its operations from intrusive interference in the second. Nothing is more important in training the mind, than the giving to it, from the first, habits of concentrated attention, and powers of sustained abstraction. It is in this double way that the will is mainly exerted in aiding the operations of the intellect.

614. The *emotions* have properly nothing whatever to do *with those high operations* of the mind, which concern themselves only with the investigation and discovery of truth.

But people in ordinary life are scarcely ever able to

distinguish between their desires, and their convictions. They take their wishes for ascertained truths, and reason out results from them ; the consequence is that those results are more often errors than realities. The next most important duty in mental training, after the formation of habits of attention, and powers of abstraction, is the learning that desires are things which are always to be treated with hesitation and doubt, instead of with confidence and trust. Men, whose minds have been badly trained, always *believe what they wish to be true*, instead of seeking truth patiently, laboriously, self-denyingly, and with extreme caution.

615. *Patience and caution* are *habits* of the mind, that are absolutely *essential to philosophical investigation.*

People are generally so impatient in their investigations, that whenever they find difficulties in their way, they jump to hasty conclusions, which are not sufficiently established by evidence. In doing this, they mostly prefer error to doubt. Nothing is more common in life than to find persons of narrow intellect and few ideas, who are ready to decide on the instant upon questions that have occupied the thoughts of the most intelligent men for years, without their being able to come to satisfactory conclusions with regard to them. In consequence of the gradual and step-by-step way in which all the deductions of intellect are made, there must be countless numbers of subjects, in all thoughtful minds, that are unsettled matters. It should be a portion of every one's training to enable him to remain patiently amidst these doubts, until they are made clear to him by the help of an enlarged understanding.

616. The intellect and reason are *developed and strengthened by training and education.*

It is in this respect that the intellect and reason differ from instinct (566.) The more carefully the mental faculties are trained, the more powerful and perfect they become. This is the necessary result of the statement

that has been already made regarding the indefinite expansiveness of mind (596.) All human life is one long
continuance of education. The only question is whether
that education is well or badly carried on.

**617. The proper object of all education is
*the replacing of instinctive impulse by intel-
ligent reason.***

It has been stated that instinct is perfect at the commencement of life (566;) hence, the infant of the human species comes into the world with strong impulses
acting upon its little frame. There is then, however,
no single idea present to its mind. Its entire being is
simply a state of pleasurable or painful consciousness,
in which it does certain things under the influence of
impressions of sense; but, at the instant of birth, its
intellectual education begins. By slow degrees sensations are converted into ideas, and these ideas are examined and reasoned upon. The results of the reasoning processes, then, get mingled more and more with
the impulses, controlling, modifying, and altering them,
as the cerebrum is progressively developed and acquires
greater power in the organization. It is because the
mind is fed by sensations changed in the cerebrum into
ideas, that education possesses influence in forming and
moulding the character. The more carefully and precisely the ideas are fixed, and the more rigorously
habits of attention, abstraction, and cautious and patient
generalization are framed, the more intelligent and rational the creature will become. The business of the
education is to exercise the cerebrum, and make it grow
by use in force and power. There is the clearest evidence that not only its strength is increased by judicious
and continued exercise, but that even its bulk is augmented. The word *education* is derived from the Latin
terms " *e*" and " *duco*," which mean to " lead or draw
out of." Education " leads out" the faculties of the
cerebrum, and gives them weight over the impulses.
The well-educated man never acts upon impulses. He

pauses upon them, and lets them produce trains of thought and reflection, instead of immediate action. Whenever this is done, the course that is ultimately pursued is almost sure to be determined by different motives from such as would have been operative at first. The object of educational training is, therefore, to make the mind vigorous and clear, and to give it power of control over the impulses.

618. Intelligence and reason have been made dependent upon education, in order that the mind *may possess unlimited powers* of growth.

It has been stated that all purely instinctive creatures, which have to fill the same position in the scheme of nature, are exactly like to each other. Rational creatures, on the other hand, are all unlike to each other; each one is marked by a distinctive character of its own. It is as impossible to find two men exactly like each other, as it is to find two bees that could be distinguished, the one from the other by its character. This is because each mind is *drawn out* in a line of its own. One mind goes in one direction, and another quite a different way; still each one can continue to *advance* in its own direction, so long as its education continues, that is, so long as it is active in itself. Uniformity of opinions and thoughts in different minds is not the result of education, but rather of mental fixation; where such occurs, there is no real *leading out* of the faculties going on. In such cases the thoughts are cast and confined in rigid moulds, instead of being led to vital growth. Education is designed to produce diversity of character, and not uniformity. No rational mind ever expects or desires to see another exactly like to itself.

619. The cerebrum and sensorium are *not always active*, like the spinal cord. *

Every change of a state of consciousness into an idea,

and every perception of an idea is attended by waste of
the substance of the cerebrum and censorium. Hence,
all mental operations tend to exhaustion of its powers.
The cerebrum, as has been before stated (581), is one of
the most active organs in the frame. In order that it
may be so, it is supplied with much more blood than
any other part of the organization ; still, even this abundant supply is not sufficient to make up the waste of its
substance. when it is actively at work. Hence, at frequent intervals, it is thrown into a state of rest, in order
that its waste may be repaired, and its structures fitted
for renewed exertion, before another call upon it is
made.

620. The rest of the cerebrum is called
sleep.

During perfect sleep, both the sensorium and cerebrum are completely inactive. The spinal cord alone,
then, continues its operations, in order that the springs
of organic life may not be stayed (555). Sleep is a state
of suspended will and absolute unconsciousness, both as
regards sensations and ideas. Soundly sleeping men are
for the time in the same condition as such of the lower
animals as are devoid of brains.

621. Rational creatures need to spend
about *one-third of their time in sleep*, for
the due recruiting of their sensorial and mental powers.

During sleep an accumulation of brain force takes
place, which is subsequently, on awakening, brought into
use and gradually exhausted. Full-grown men generally require about seven or eight hours' sleep in every
twenty-four; but the exact amount needed depends
very much on the use that is made of the mental
faculties during the waking hours. In a general way,
those who sustain the greatest amount of mental labor,
require the greatest quantity of sleep for their due refreshment. Sleep is, however, sometimes more sound
and refreshing than it is at others. One hour of sound

sleep refreshes more than two hours of that which is less sound.

622. *Dreams* indicate that the sleep is of an *imperfect* and unrefreshing kind.

During dreams, the sensorium is unconscious of the impressions of sensation, but is partially conscious to those of ideas,—that is to say, it is asleep to sensations, but awake to thoughts. The cerebrum, too, is partially active—trains of ideas follow each other by suggestion and association, but the will is perfectly powerless to direct or control them. Hence, the thoughts take the most absurd and irrational directions. In dreaming states, an impression on an organ of sensation is transferred directly to the cerebrum, and awakens there an idea without producing any trace of consciousness as it passes through the sensorium, but the idea then returns to the sensorium, and calls up perceptive consciousness in it. The ideas follow each other in dreaming with inconceivable rapidity ; each suggests another as fast as it is itself perceived. There are cases on record, in which a sound has led to a dream, consisting of a long course of events, apparently spread through weeks, and which has, nevertheless, awakened the sleeper on the instant. In these cases, the long dream must have been passed through in an instant of time.

623. The power of the will over the direction of the ideas *is sometimes suspended without the presence of even imperfect sleep.*

The state known as "reverie" is of this kind. Idea succeeds to idea through suggestion, without the will being able in any way to influence the result. The curious conditions produced in the exhibitions of what is called electro-biology are of a similar nature. The subject is in a state of reverie, and is for the time separated from all connection with his own will. His movements are all directed by ideas that have taken strong possession of his mind, and cannot be replaced by any effort of his own. The great peculiarity of the condi-

tion, as contrasted with ordinary reverie, is that those
ruling ideas are called into activity by the conversation
of the exhibitor. The will of the exhibitor has no other
power over the subject than that he is able to call up
the ruling idea through strong association. He suggests
the idea by his words, and forthwith the idea becomes a
state of consciousness, and determines the actions. As
in the states of instinct, the actions are all produced
directly by sensations—in the conditions of so-called
electro-biology, the movements are produced *directly by
ideas;* in both cases the result is irrespective of the influence of voluntary control.

CHAPTER XX.

THE EXTERNAL SENSES.

624. Animals acquire a knowledge of external objects through the *instrumentality of
their senses.*

Feeling and consciousness are really changes in the
mind, produced through impressions effected upon some
part or other of the body. These impressions are made,
as has been already seen (540), upon the external extremities of nerve tubes, which are particularly fitted for
the conveyance inwards to the sensorium, of some influence rendered active by the impression. But there
are different kinds of feelings produced in this way.
Some give information concerning one kind of quality in
external bodies; some concerning other kinds. Nerves
of sensation are therefore suited for the reception of different kinds of impressions in different parts of the body.

They are grouped together at their outer extremities, as different kinds of instruments or organs, modified according to the work they have to perform. These different kinds of feeling instruments are called " *the organs of the senses,*" or, more familiarly, the "*external senses.*" (The word "*senses*" is derived from "*sensus,*" a Latin term, which means, amongst other things, the organ through which external objects are felt. This is primarily taken from "*sentio,*" to "*feel.*")

625. The sense which gives information relating to resistence and hardness is called *the sense of touch.*

When the finger is pressed against a table or a heavy book, the knowledge is acquired that some material substance occupies the position in which the resistance is felt. Some notion is also gained regarding the hardness or softness of the resisting body. The way in which a wooden table withstands the finger is distinguished from the way in which a paper book does the same thing. All ideas of material substances being external bodies in regard to the perceiving individual, are procured through the sense of touch. It reveals the existence of an external world. Some of the simplest kinds of animals have no other sense than this of touch. They have no other means of learning anything about external objects than by resting in contact with them, and so receiving an impression from their forms. (The word "*touch*" is derived from the French "*toucher,*" "*to handle.*")

626. The sense of touch is exercised by the *entire external surface* of the body.

It is by its outer surface that the body must be brought in connection with external objects. Hence, the nerves that convey inwards the impressions of touch, have their extremities scattered through the skin. The outer surface of the skin is covered by several layers of dead dry cells or scales, arranged over each other like the tiles of a house, and together forming what is called the scarf skin (753). Beneath this a vast number of little points or papillæ (*nipples*) project, as shown in *Figure* 72. These papillæ of the skin are made of common fibrous tissue (437). They are about the one hundredth of an inch long, and one quarter as broad. A loop of the capillary blood-vessels is furnished to each to supply it with nourishment; but a nervous tubule also ends in it (537). These papillæ of the skin are the parts in which its powers of feeling are situated. They are indeed the organs of touch. Whenever the hands are pressed upon hard bodies, the points of these papillæ are squeezed in, and as a consequence of the pressure, some nervous influence is produced, and sent in to the sensorium along the nerve tubule. Very

Fig. 72.

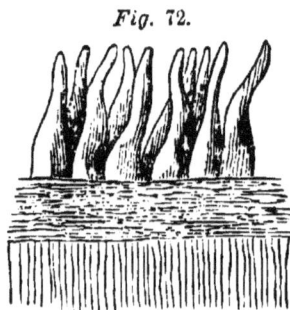

curious cellular bodies have been found to be connected in great numbers with the terminations of the nerves of feeling that lie in the skin, which in all probability have something to do with the origination of the influence that is conveyed by the tubules (541). These are called the Pacinian bodies, because they have been carefully examined and described by Professor Pacini of Pisa. The papillæ of the skin are most numerous and most thickly planted in parts that are most sensitive, as in the hands and lips.

627. The nerves of general sensation are capable *of conveying certain other impressions* besides those of touch.

It is through these same nerve tubules, which carry information of impressions of touch, that sensations of warmth and cold, of tickling and itching, of pain, and of various other irregular kinds, are conveyed to the sensorium. Nerves, corresponding with those of general feeling that run from the skin, are transmitted from all the internal organs of the body, and carry sensations of hunger, thirst, sickness, fatigue, and of other states and impressions too numerous to be mentioned. These are all modifications of the simple mechanical impression of touch. All these sensitive *afferent nerves* (535) from the skin and internal organs meet together with the motor or *efferent nerves* (538), and help to constitute the several pairs of spinal nerves that are in communication with the different portions of the frame (517). They all then run up through the back column of the spinal cord (550) to the sensorium, and terminate in connection with its vesicular mass.

628. Pain is merely common sensation *raised to a very intense degree.*

Sensation of every kind is productive of pleasure, so long as it is of a nature that is wholesome for the organization of the body. When it is not wholesome for this organization it becomes painful, and so impels the creature to exert its powers of movement, and to attempt the removal of its frame from the injurious influence. All very strong impressions are dangerous to the delicate organization that receives and conveys them. Hence, intense sensations are made to be painful. Pain is the great guardian of the well-being of the frame. When a finger is thrust into the fire, it will be destroyed in a few instants if allowed to remain there; but the pain of the burn, the instant it is experienced, causes the organ to be drawn back, and so saves it from destruction. Whenever pain is felt, something is wrong that needs to be remedied. In most cases nature herself supplies the remedy, as when people suffering the pain of internal inflammation, go to bed and keep themselves quiet until their sense of illness and discomfort is removed. Sometimes, however, ignorance on this point, or wilfulness, leads to much harm, nature being crossed and opposed in her beneficent intentions, instead of being helped.

629. The sense which gives information re-

garding the nature of the food that is taken, is called the *sense of taste.*

In a general way, those substances have an agreeable taste that are safe and suitable to be employed as food; but those that are poisonous or unsuitable have disagreeable flavors. The sense of taste is given to animals that they may *try their food* before they swallow it. (The word "*taste*" is derived from "*test*," the French term for "*trial*.") Animals make trial of what they taste, and instinctively decide upon its being suitable or unsuitable to be employed as food, accordingly as they derive pleasure or annoyance from the trial.

630. The sense of taste *is exercised by the tongue.*

The tongue is aided in an inferior degree in its task of making trial of the food by the membranes of the mouth. The mouth, into which food about to be swallowed is first taken, is obviously the situation in which the organ should be placed that is destined to make trial of the food. The tongue and membranes of the mouth are furnished with a quantity of little points or papillæ of a very similar nature to those of the skin (626). In these papillæ loops of capillary blood-vessels, and terminations of nerve tubules are contained. The nerve tubules are gathered together into main branches which constitute the nerve of taste. These are sent up from the back of the skull, through holes in its floor, into the upward prolongation of the spinal cord (557), with which they pass to the sensorium.

631. Substances that are tasted are dissolved *in the fluids of the mouth,* and so brought into very close communication with the papillæ of the tongue.

The papillæ of the tongue are made upon the same general plan as those of the skin. The organ of taste is indeed a modification of that of touch; but it is also a refinement upon it. It has much more delicate work to perform. It has to distinguish a very great many substances from each other, and often by shades of difference that no delicacy of touch could detect. The papillæ are covered by the living cells of mucous membrane (442) instead of by the dry and dead scales of the scarf-skin (626). These cells pour out liquid mucus (445) which, with the help of the saliva, dissolves portions of the food that is under process of chewing in the mouth. The solutions are conveyed into the cells by osmose (163), and so are placed in close communication with the papillæ and their nerve-tubules that lie beneath. The peculiar influences, which they originate in the extremities of the nerve tubules, then get transmitted into the sensorium along the branches of the nerve, and are there distinguished from each other, if different in ever so slight a degree. Some produce perceptions of pleasure, and some perceptions of discomfort or pain. The close resemblance of the sense of taste to that of touch is shown in the fact, that the actual contact of the organ with the substance to be tasted is essential to the exercise of the sense. So also the part of the tongue which is the seat of direct sensation can be distinguished, just as the part of the skin that feels an object can. The nerve of taste, too, joins the spinal cord (or its upper prolongation) before it is connected with the sensorium, just as the nerves of common sensation do.

632. The sense that gives information concerning the nature of odors that are floating in the air is called the *sense of smelling.*

The word "*smell*" has been in use to express this power for a very long time. But it is not now known why the term was first employed for the purpose. (It · has been imagined that it may have been derived from the Anglo-Saxon "*smœccan*," "*to smack.*"

633. The sense of smelling is exercised by the membranes spread out *in the cavity of the nose.*

L

The internal cavities of the nose constitute the organ

of smelling. *Fig. 73* shows how these cavities are arranged. It is a sketch of the inside of the nose seen from behind *A*. Partitions and scrolls of bone project into a hollow space which opens behind into the throat, and before, by means of the nostrils, into the external air. This cavity is separated from the mouth, *M*, by the floor or arched palate of bone, *P ;* and from the interior of the skull. *S*, by a thin plate of bone pierced full of small holes. The sides of the cavities are everywhere covered with mucous membrane (443), in which the extremities of a great number of nerve tubules are spread out. These tubules are gathered together

Fig. 73.

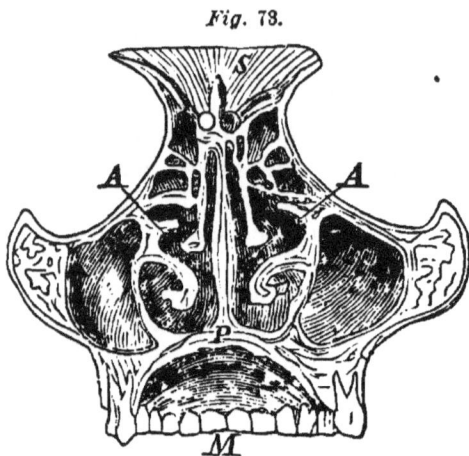

Fig. 74.

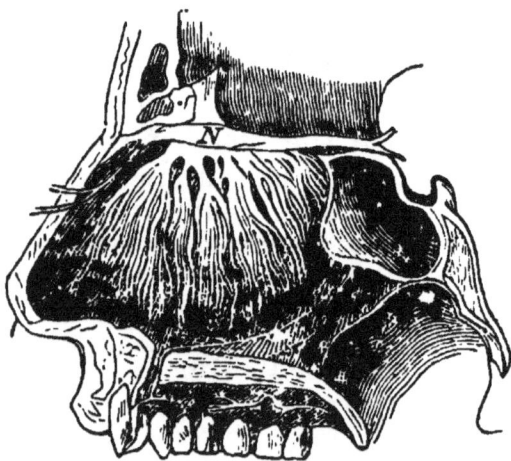

into branches, which pass up through the holes in the
ceiling of the cavities, become the nerves of smelling in
the skull, and are there joined to the sensorium. *Figure
74* represents the immense number of tubules of the nerve
of smelling that are distributed to the membrane that
lines the cavity of the nose, which cavity is here seen
sideways. *N* is the common internal termination of
these tubules, where they are in connection with the
sensorium.

634. The sense of smelling is excited by
*particles of odorous bodies being placed in
contact with the membrane* that lines the
cavity of the nose.

All bodies that are odorous are volatile, more or less:
that is to say, are capable of sending out extremely
minute particles of their substance floating into the air,
as water sends out vapor (79). These particles are
drawn through the nostrils, with the air, in the act of
breathing, and so come into contact with the moist mucous membranes of the cavity of the nose, and get to
affect the nerve tubules that are in connection with them.

635. The sense of smelling *determines the
quality of the air* that is breathed.

The sense of smelling does for the chest what the
sense of taste does for the stomach. It warns against
the admission of improper or dangerous materials. In
a general way all volatile substances, that are disagreeable to the organ of smelling, are injurious, if admitted
into the chest, and, on the other hand, all that are agreeable are safe. The nose is, however, an inlet of information, as well as a guard against evil. It helps to distinguish different bodies from each other, and so is
employed by intelligent creatures as an auxiliary in collecting materials for the formation of ideas. Most of
the lower kinds of animals are destitute of the sense of
smell. Its possession is somewhat of a lordly privilege.

636. The sense that gives information con-

cerning the *nature of sounds* is called the
sense of hearing.

The word "*to hear*" is a modification of an old Anglo-
Saxon term, "*hyran.*"

637. The sense of hearing is exercised by
a pair of ingeniously contrived organs, called
the ears.

The ears are the organs of hearing. The organs of
touch, taste, and smell, are little more than modifications
of the general surface of the body. They are merely the
intermingling of sentient nerve tubules with the covering
or lining membranes of the frame, so that these membranes may be made sensible of the contact of bodies,
and be able to distinguish differences in their natures or
characters. The ear has, however, a far more delicate
and important task to perform. Hence, it is a very complex organ, built up with great ingenuity and care.

638. Sounds, which are the objects on
which the sense of hearing is exercised, *are
produced by vibrations amidst the particles*
of material bodies.

Nothing sounds unless it vibrates. Vibration means
a beating backwards and forwards, or trembling, of the
particles of any body. (The word is derived from the
Latin "*vibro,*" which signifies "to tremble"). Any one
may observe that the waves of the sea beat successively
upon the face of a rock forming part of an abrupt sea-
shore. The only difference between waving and vibrating is, that in the latter case the same action goes on
throughout the substance of the body, which in the former
case takes place only on its surface. When anything is
made to vibrate, its entire substance trembles like the
surface of the sea. Sound is the result of this vibration.
If the rim of a large drinking glass is struck smartly, it
emits a clear sound like the tone of a bell. Let the glass
be poured full of clear water, and then be struck in this
way, and the cause of the sound will be at once apparent.

So long as the glass is sounding, the surface of the water will be seen to be agitated by a series of little ripples. These are disturbances communicated to the water by the tremblings of the glass.

639. Vibration in the substances of material bodies *results from the property known as elasticity.*

The word "*elasticity*" is derived from the Greek term "*elao*" which means to "*shake*" or "*drive.*" "Elasticity" is used to express the property by means of which any body tends to recover its original size and form, after these have been altered. A piece of india-rubber that has been pulled out, springs back to its original length when let go, because it is elastic. Now when any material body is suddenly and sharply struck, the particles upon which the blow falls are driven before it; but they are then forced upon their neighbors which at first resist them, and then drive them back. As they return to their original position, they acquire force like an unloosed spring, and so overshoot their mark. Having done this for a little way, they are drawn back yet again, and the same series of actions is renewed, and continues for some little time growing fainter and fainter, until at length the vibratory movement subsides into rest. The particles that are struck by the blow, in their turn strike their neighbors, and so produce a like effect on them. In this way vibratory trembling is made to pass through the substance of bodies, in consequence of their elasticity.

640. Sound is *conveyed to the ear by the air.*

Vibrating bodies are every where closed in by continuous masses of air, which is itself an elastic body (61, 62), and which therefore vibrates when disturbed. When the rim of a drinking glass is struck, it agitates the air in contact with it, just as it does water when in the same position. But the tremblings in the aerial substance then run through very great distances, and if it chances that an ear lies in their way, they beat upon it, somewhat as the waves of the sea beat upon a rocky shore. When the tremblings of vibrating air beat upon an ear, *sound is heard.* Sound is really a peculiar state in a series of living nerves, that are in connection with the sensorium, and is produced by the beating upon their extremities of little vibrations, these vibrations having been carried from the bodies in which they first began, by the elastic air that fills up all otherwise unoccupied spaces upon the surface of the earth. There is thus this peculiarity in the sense of hearing to distinguish it from the senses—touch, smelling, and taste. The objects that excite its sensations, are never in contact with the organ. The influences which they originate are sent to it across wide tracts of distance by an intervening medium, the air. In the absence of air, no sound is conveyed to the ear, however much bodies may be made to vibrate and tremble. When a bell is struck by its clapper under the receiver of an air-pump, from which all the air has been pumped out, no sound is heard.

641. The vibrations of trembling air first *enter the trumpet-like tube* of the ear, when they produce sound.

The external ear is formed of a broad hollow plate of gristle and skin, which leads to a narrow curved tube, rather more than an inch long. This tube runs into the substance of one of the side bones of the head. In *Figure* 75, the general course and direction of this tube is shown, as if sliced through and viewed from the side. *E* is part of the gristle of the external ear. *T* is the curved tube to which its trumpet-shaped cavity leads. *B B* are parts of the thick bone of the skull into which the tube runs,

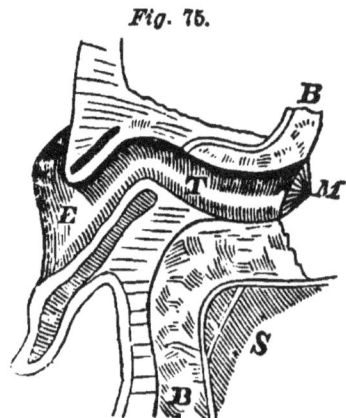

Fig. 75.

and *S* is the inside of the cavity of the skull where the brain lies.

642. The inner end of the external tube of the ear *is closed by means of a membrane stretched across it, as the parchment is* stretched across the end of a drum.

The position of this membrane, which closes the internal extremity of the tube of the ear, is shown at *M* in *Figure* 75.

643 Inside of the membrane which closes the tube of the ear there is a cavity or chamber filled with air. This chamber *is called the drum of the ear.*

This cavity is hollowed out in the substance of the bone which is here made particularly thick and strong for the purpose. A portion of the thick bone in which the internal part of the organ of hearing is hollowed out, may be felt directly behind the ear. The air-filled cavity is called the drum (or, in Latin, the *Tympanum*), from a fancy the old anatomists had to liken it with its outer membrane, to the musical instrument known by the same name. The membrane is termed the membrane of the drum. In *Figure* 76, *D* represents the cavity of the drum of the ear, hollowed out in the substance of the bone, *M* being the inside of the membrane that divides it from the outer tube. (*M also in Figure* 75.)

Fig. 76.

644. On the *inner wall* of the cavity of the drum (opposite to its outer membrane)

there *is another opening* closed with a plate of membrane.

This inner opening with its membrane is called the "oval entrance or window" (*fenestra ovalis* in Latin.) The position of the inner membrane of the tympanum or "oval entrance," is shown at *f* in *Figure* 76.

645. The outer and inner membranes of the drum of the ear are *connected together by a chain* of four little bones.

These little bones of the ear are sketched at *b b* in *Figure* 76. Two, shaped respectively like a hammer and anvil, are attached to the outer membrane, and the bottom of one, shaped like a stirrup, is firmly fixed over the inner membrane (at *f*). They are all connected, too, with each other, and thus stretch as a sort of chain across the cavity of the drum. In this way, vibrations are carried across this middle chamber of the ear. When the waves of the air beat upon the outer membrane of the drum, it is made to tremble. When it trembles, the little chain of bones connected with it does the same thing, and when the bones tremble, so also does the membrane of the "oval entrance" at their inner termination.

646. The inner membrane of the drum leads to an *internal chamber*, hollowed out in the bone beyond, and called *the labyrinth*.

Fig. 77.

The oval entrance is a door, rather than a window, and leads to the labyrinth of the ear. This innermost chamber of the ear is called the labyrinth, on account of its intricate shape. (*Labyrinth* is derived from the Greek " *labiros*," an intricate cave.) *Figure* 77 shows the form of the labyrinth of the ear, supposed to be seen from without. It consists of three semicircular canals, *c*, and one spiral one, *s*, ending in a point, all opening out from a vestibule or entrance chamber, *v*; *f* is the inner membrane of the drum, which separates the cavity of that middle chamber from the cavity of this inner one (*f* also in *Figure* 76).

647. The labyrinth of the ear is *lined by a membrane*, on which the extremities of the tubules of the nerve of hearing are spread out. In this way the inner surfaces of these bony canals are made sensitive, so that they can feel any tremblings communicated to them. The nerve tubules spread out upon them have nerve cells (541) and capillary blood-vessels associated with them, and they are collected together into branches, which run into the cavity of the skull through a hole left in the bony plates for the purpose. There they become the nerves of hearing, and join the sensorium.

648. The canals of the labyrinth are *filled with a liquid* like water. The chamber of the labyrinth is entirely filled with fluid. The consequence is, that when the inner membrane of the drum is set trembling by the little chain of bones, it makes the liquid of the labyrinth, in contact with its inner side, tremble in the same way. Thus the tremblings of the outer air are conveyed to the extremities of the nerves, spread out to receive them on the walls of the inmost recess of the organ of hearing. All the following processes must take place before a sound can be heard:—1. The sounding body must be made to vibrate or tremble. 2. The vibrations must be carried and communicated to the membrane of the drum of the ear by trembling air. 3. The membrane of the drum must make the chain of bones that crosses the middle chamber of the ear tremble. 4. This chain must cause

18   L*

the inner membrane of the drum to tremble.   5. The
inner membrane must communicate its tremblings to
the liquid contained in the channels and canals of the
innermost chamber or labyrinth.   And 6. The nervous
lining of these channels must feel the tremblings of the
liquid, and communicate them through the nerve tubules
to the sensorium, the seat of all sensation.

649. The nerve of hearing *distinguishes
the character of the vibrations* that are communicated to it through the liquid of the
labyrinth.

The tremblings that are communicated to the nerve
tubules through the apparatus comprised within this
wonderful little organ, may be strong, or they may be
weak, or they may be modified in a great number of other
ways. All these differences are at once distinguished from
each other when the impressions are communicated to the
sensorium, and are marked as separate sounds. It has been
ascertained, that some audible sounds make as many as
twenty-four thousand beats upon .the drum membrane
of the ear every second ; on the other hand, sounds are
heard under peculiar circumstances, that make only
eight beats in the same time. The frequency of the
beats on the ear determines the high or low tone of the
sound, the most rapid vibrations producing the highest
tones. But the beats may also be more or less intense
without being more or less frequent ; and so also they
may vary in breadth, and in the manner in which they
fall on the membrane. All these variations alter the
quality of sounds, without affecting their tone. The
organ of hearing is able to detect the slightest differences in all these several conditions, when the beats of
the vibrations fall upon its nerves.

650. The ear is thus an apparatus fitted to
collect the tremblings of the air, and *throw*
them upon a *great number of nerve tubules
at* once.

There are an immense number of nerve tubules distributed upon the internal membranes of the ear; these are all concerned in feeling the same vibrations at the same instant of time. The number of tubules employed in the investigation, probably enables the work of distinction to be more exactly and delicately performed. Their extremities are all stretched out, as if to catch and register every shade of variation in the tremblings of the liquid lying in contact with them. It will be at once understood, that the work they perform is of the utmost delicacy, and needs the exertion of the utmost nicety of discrimination, when it is remembered that they distinguish with never-failing accuracy, not only musical notes running through several octaves, but also all the refined differences of articulation and expression employed in speech. It is through the organ of hearing that articulate words are made so full of instruction and meaning (697). On account of its relation to language, the ear is one of the most valuable of the inlets through which the materials of ideas are admitted into the mind.

CHAPTER XXI.

THE EXTERNAL SENSES—SENSE OF SIGHT.

651. The sense that gives information concerning the *form and appearance* of external objects is called the sense of sight.

The word sight is derived from "*sigh*" (*sigh'd*), the old form of the past tense of " *to see*"—Anglo-Saxon, " *seon.*"

652. The sense of sight is exercised by a pair of complex organs, called *the eyes.*

The eyes are the organs of sight. (The word "*eye*" is derived from the Anglo-Saxon "*Eag*," "*augyen,*" "*to show.*") The eyes, like the ears, are complicated pieces of apparatus, arranged and fashioned with a most wonderful adaptation to work that has to be performed.

653. Surrounding objects *are made visible by the presence of light.*

If the shutters of the windows of any room filled with visible objects, but having neither fire nor other artificial illumination in it, be suddenly closed during the day-time, all the several objects disappear at once, just as if they were swept away. The shutting out of the light makes the objects cease to be seen, consequently it was the light that in some way made them visible before.

654. Light is produced by the *vibrations of a very elastic medium,* much in the same way as sound is.

"*Medium*" is a Latin word, signifying any thing that comes between, or is common to several objects. It has been adopted into the English language by men of science, to express the presence or existence of any thing that occupies space. Thus air is a medium which occupies the spaces that lie between terrestrial objects. Sound, therefore, is the trembling of the elastic medium, known as air. But men of science have found reason for the conclusion, that all the wide spaces of the universe are filled with a medium that is very much more subtle, and very much finer than air. All the worlds and suns of the universe have a thin medium filling the spaces that lie between them, as air fills the spaces that lie between terrestrial bodies. This universal medium is elastic, and its tremblings constitute light, just as the tremblings of the air constitute sound. Air does not stretch between the several worlds of space; it is confined round each as a limited atmosphere (64). Hence, sounds cannot be sent through intervening space from one world to any of the rest. But the universal medium now under consideration does stretch from body to body, and sun to sun. Hence, one of these bodies can send light across intervening space to its neighbors. And hence, when, at night, man looks out to the universe that surrounds him, he sees the countless hosts of stars that are scattered through its immensity. He sees all those stars by light that has trembled through the elastic medium, which lies in the spaces intervening between them and himself.

655. **The medium that vibrates or trembles when light is produced** *is called ether*.

"*Ether*" has been chosen as an appropriate name for the subtle medium that pervades the universe. (It is the Latin denomination of the "*sky*," or "*firmament*," and is primarily derived from "*aitho*," the Greek term for "*to burn*," or "*shine*.") Nothing is known of the real nature of this subtle, all-pervading substance. It is quite inappreciable to the human senses as an object of direct observation. Its existence is only inferred from certain results that cannot be explained or understood without it, but that can be both explained and understood when its presence is taken for granted. The incapability of the human senses to detect it by direct observation, is by no means an argument of weight against its existence; for there was a time when men of science were in the same state with regard to the air. They did not see it, and therefore took no account of it as an existing body. But modern research has been able both to weigh (31) and to separate it into its elements (32). The universal ether of space is, however, infinitely more subtle than air. It seems to be actually without ascertainable weight, and its tremblings travel two hundred thousand miles before the vibrations of air can travel two thousand feet.

656. Light makes bodies visible when it falls on them by being *reflected, or thrown back from the eye.*

Visible bodies do with the subtle ether what sounding bodies do with air. They throw off vibrations or tremblings from their surface. These tremblings flash in direct lines all round, and when any of them chance to fall upon an eye in their course, they excite in it a sensation of sight. Nothing is known of the means by which visible bodies excite tremblings in the surrounding ether. It is only certain that bodies, which produce this result, must either be what is called luminous in their own natures, or they must be so placed that they can receive vibrations that have issued from luminous bodies; they then send these back second-hand from their own surfaces.

657. Light *passes through some kinds of bodies* more or less freely.

Such bodies are then called *transparent*, because objects appear through them (*trans* and *appareo*). Air and glass are good illustrations of transparent bodies. It is not known why some bodies are transparent in this way, and others not; but it is believed that the subtle ether, that vibrates to constitute light, really pervades their substance, and exists freely amongst their pores, so that its tremblings can pass uninterruptedly through them.

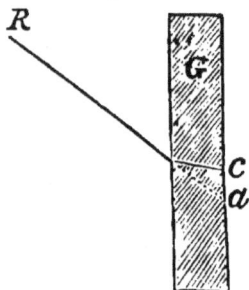

Fig. 78.

658. Light is bent into a new path when it passes *obliquely from one transparent medium into another* of a different density.

Thus glass is a denser medium than air. It has more particles than air in a given bulk. Let *R* in *Fig.* 78 represent a ray, or line of light vibrations, passing through air and falling obliquely upon *G*, a plate of glass. The ray, after it has passed into the glass, does not go on in the same straight line to *a*, but it is bent up by the glass so that it reaches *c* instead.

659. Every point of the surface of a visible body *sends off from itself rays or lines of vibrations* in straight directions *all round.*

Thus, in *Figure* 79, the point 1 in the flame of a candle scatters rays from itself in all directions, as shown by the radiating lines; but the point 2 does precisely the same thing; and so indeed do the hundreds of other points of which the flame of the candle is composed. Any flat surface, therefore, (as *s, s.*) receives vibrations from all the points confusedly mixed together. It is lit up with a confused mass of light, and is excited to send out its own diverging lines of vibrations from each of its own points, second-hand.

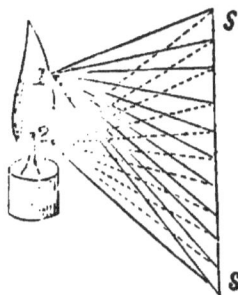

Fig. 79

660. A piece of glass, shaped like a lens, *bends these lines of vibrations*, when they pass through it, in such a way, that all those which have *issued from one point* are again brought back beyond the glass to *one point.*

A "lens" is a piece of glass curved on both sides something like the seed of the lentil. ("*Lens*" is the Latin name for this seed.) *Figure* 80 shows what happens when the several diverging rays from the flame of a candle are allowed to fall upon the surface of a lens of glass, instead of upon a flat surface, as in the preceding figure. All

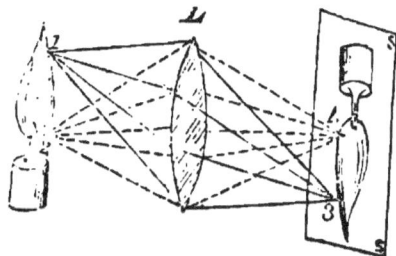

Fig. 80.

the rays that have diverged from 1 are, by the power
of the lens, drawn together again at 3.   All those
from 2 are drawn to 4; and so with the diverging
rays from all the other points of the flame.   The lens
of glass (*L*) so sifts and sorts the rays which pass
through it, that each system or series coming from a
common point is sent on towards a common point.   It
does this in virtue of the property named in paragraph
658.   Every ray falls on the curve of the lens with a
different slope or inclination from its neighbor, and is
there bent in a different degree as it passes through the
glass; but these differences are all so arranged that the
result is the sorting named above.   If now a white surface, like a card, be placed at *s s*, it will be seen that a
perfect image of the candle is there formed. only turned
upside down.   This result may be practically shown by
a very simple apparatus.   Place a bright candle on a
table; hold a flat card opposite to it, a few inches off,
and then insert a common pocket magnifying glass, or a
glass taken out of the tube of a telescope, midway between, and immediately an inverted picture of the flame
of the candle will be seen drawn in light upon the card.
(The lens will need to be moved backwards and forwards
between the candle and the card, until the image becomes distinct.)

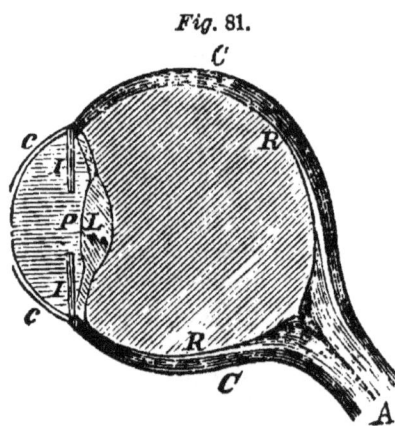

Fig. 81.

661. The eye con-
sists of a *hollow globe
or ball*, with a trans-
parent opening or
window in front.
The walls or shell of
the globe are formed of
coats of strong and tough
fibrous membrane (437);
but these coats are left
transparent in front.   In
Figure 81, the coats of
the eyeball are repre-

sented at *C C*. The front transparent portion of them
extends from *c* to *c*, and bulges out more than the rest.
It is called the *cornea* of the eye. (The name is taken
from the Latin word "*corneus*," which signifies horny.
The substance looks like a curved plate of transparent
horn.)

662. A *lens of crystal* is placed directly
behind the transparent opening or window
of the eyeball.

The position of this lens is shown at *L*, in *Figure* 81.
All that portion of the eyeball which is in front of the
crystal lens, is filled with a thin liquid, like water, which
is called the "*aqueous*," or watery humor of the eye.
All that portion which is behind the crystal lens is filled
with a thicker humor, which is called the "*vitreous*"
humor (from "*vitreus*," for Latin "*glassy*"). The body
of the eyeball is thus composed of three transparent
media of different degrees of density, one like water,
another like glass, and another like crystal. The crystal
lens has the greatest power to bend and sort rays of light
passing through it, but the other humors help it in its
work, and correct some of its imperfections.

663. The crystal lens of the eye *acts upon
light vibrations*, when passing through it,
exactly in the same way *as a lens of glass
does.*

So that if the eyeball were placed in the position
represented by the lens of glass in *Figure* 80, with its
transparent window looking towards the flame, the picture of the flame would be drawn in light upon the back
of its inside at *R R*, (*Figure* 81,) just as in the other
case it was drawn upon the card *S S* (*Figure* 80).

664. An immense number of *nerve tubules
are spread out at the back of the eye*, where
pictures are formed by the instrumentality
of the lens and transparent humors.

Thus, in the eye, the picture is formed on a living surface that is capable of feeling it, instead of merely upon a card-board surface that can only receive and reflect it, as in the experiment described in paragraph 660.

665. The nerve tubules of the back of the eye are *collected together into a large nerve*, which runs into the cavity of the skull to join the sensorium.

The beginning of this nerve of sight is shown at *N* in *Figure* 81, running off from the back of the eyeball on its way towards the brain.

666. The nerve tubules, spread out on the back of the eyeball, *are arranged as a sort of net-work*.

Hence, this inner nervous lining of the eye is called the *retina*, *R R*, *Fig.* 81. ("*Retina*" is the Latin term for "*a net.*") In the retina there are, however, a couple of layers of nerve cells, which are intimately connected with the terminations of the tubules (541); and behind it is a layer of capillary blood-vessels, destined to supply the material support of the organ's activity (543). *Figure* 82 shows the arrangement of these parts in a portion of the internal nervous lining of the eyeball. *V V*, *V V*, are 2 layers of vesicular nerve-matter gray with meshes of fine blood-vessels. *R R* is the network of nerve tubules placed between them. *M M* is a membrane composed of columnar rows of cells, like the pile of velvet, which is placed outside of the nervous layers as a protection to them. This membrane is covered externally by a coating of a sort of black paint, which effectually stops the vibrations of light from passing on any further, and so allows all their force to be

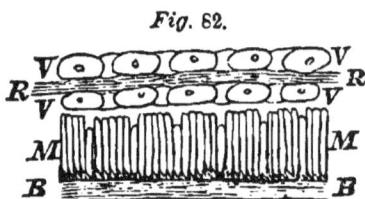

Fig. 82.

expended in producing sensory impressions in the nervous substance. Beyond this dark film lies *B B*—a coat composed entirely of blood-vessels, and from which the minute branches are sent to *V V*, the vesicular layers. Over this lies the outermost, strong, fibrous coat of the eyeball.

667. The nerve tubules of the eye *distinguish different kinds* of tremblings, that are communicated to them through the transparent humors of the organ.

In this respect they resemble the nerve tubules of the ear. The frequency of the vibrations determines the color that is perceived, and their intensity, the amount of light or shade with which the object is invested. The perception of a violet color is produced by rapid vibrations of ether striking on the nerve fibrils, as high tones are produced by rapid vibrations of air striking on the membrane of the ear. The perception of a red color by comparatively slow vibrations of ether striking the eye, as low tones by slow vibrations of air striking the ear. Orange, yellow, green, and blue, stand between red and violet in the frequency of their vibrations. Sir Isaac Newton has calculated that 421 *billions of vibrations* strike on the nerves of the eye every second when the sensation of red light is experienced, and 799 *billions* when the sensation of violet light is produced.

668. Each nerve tubule of the eye receives and conveys *a distinct impression from its neighbor.*

In this respect the eye differs from the ear. All the host of nerve tubules in the ear are busied in receiving the same impression at the same instant of time, and the result is the perception of only one sensation or sound at one instant. But each nerve tubule, of the host in the eye, is working alone, receiving and transmitting its own impression. A great many impressions are received at the same instant by the eye, and are all arranged in their proper relative positions; each one being marked by its own intensity and color, according to the manner in which it has been thrown off from its own point on the surface of the object that is made visible by the impression. It is by this formation of a light picture on the nervous lining of the eye, in which picture each point has its own intensity and kind of impression, and its own separate nerve tubule placed beneath it to receive and distinguish these characters of the impression, that vision is performed. Vision gives direct information concerning the outlines and shapes, the lights and shades, the colors and relative positions of surrounding objects. All these particulars are mapped accurately down in the eye, where they can be felt, and the mapping is performed by a host of little messengers —the tremblings of light—which are sent off from the several points of those objects for the purpose of making an exact register in the proper place.

669. The apparent size of an object depends upon the *dimensions of the picture* that is formed as its image on the nervous lining of the eye.

When bodies are brought nearer they seem to grow larger, because their pictures formed within the eye are then made to cover a larger portion of the retina. The same thing happens when bodies are viewed through magnifying glasses. Their pictures are made to spread themselves out more. When a great number of nerve tubules are engaged in feeling the details of a pictorial impression, they of course effect the work more minutely and more delicately than when fewer are engaged on the task. Hence, details are seen in magnified bodies that could not be made out before they were enlarged.

670. The nerve tubules that receive the separate impressions in the eye *are almost inconceivably small.*

This will be at once inferred when it is remembered

how small the globe of the eye is, compared with the vast landscape that can have its picture drawn within its chamber. All the details of the largest landscape are painted in the eye picture with the utmost clearness. Yet each point in the detail that can be distinctly seen, must have its own nerve tubule to itself in the act of perceiving it. The picture of the full moon, that is formed in the eye when that luminous body is looked at, is so small that 53 thousand such might be painted on a square inch of surface. Yet specks of shade only one-tenth the width of the moon can be distinctly seen upon its face. Of such specks five millions could be drawn on a square inch. The eye of a man can be distinguished when he is standing 40 feet away. The picture of that feature, drawn upon the retina of the person looking at him, is then so small that four millions of such pictures could be made in a square inch. Bodies can be seen that are in themselves only the five-hundredth of an inch across, and lines that are only the five-thousandth of an inch across. The pictures of these, formed on the retina when they are seen, must be many times smaller than the objects themselves, yet each point that is distinctly perceived, must have its own nerve tubule for the performance of the work.

671. Light that is *neither too weak nor too strong* is best for the purposes of distinct vision.

The nervous apparatus of the eye is of so delicate a nature that very intense vibrations are dangerous to it. On the other hand, the vibrations emitted from bodies may be too faint to have power enough to excite vision in the nerves. On this account a movable curtain has been placed behind the transparent window of the eye, in such a way that the opening of the window may be diminished when the light is too strong, or enlarged when it is too weak. The position of this movable curtain is shown at *I*, in *Figure* 81. It is called the *Iris* of the eye, and the opening in its middle (*P*) is

called the *pupil.* When the eye of any person is looked
at in front, the pupil is seen as a dark black circle in
the centre. Round this is a colored band, which is the
iris, and which may be noticed to be constantly contracting and dilating, if the strength of the light that
falls upon it is varied. The white of the eye which
surrounds the iris is a part of the outer membranous
coat. In front of the iris and pupil, the transparent
cornea bulges out. It can hardly be seen on account
of its transparency. (The word "*iris*" was the Latin
name for a "*rainbow.*" The colored curtain of the eye
has been so called on account of the brilliancy and
variety of its tints in different people. "*Pupil*" is derived from "*pupilla,*" which was the term by which the
Latins designated this part.)

672. The *two eyes* see only *one picture.*

Each eye has a picture of its own; but these two
pictures are combined together by an act of the mind,
and so are seen as one. The separate impressions
transmitted to the different eyes, are registered together
in the sensorium, and so mingled together. Each eye
has a different picture of any *near object* painted upon
its retina, from that which its neighbor has. The combination of these two *unlike pictures* by the mind confers the notion of solidity upon the object. Hence,
things are seen standing out in bold relief, and not
merely as if forming a part of a flat picture. If two
pictures are drawn of some one object, each being exactly what each eye sees when the object is near, and if
these two pictures be then viewed at once by the two
eyes, through prisms of glass that have the power of
blending them together and making them look like one
picture, that one picture has all the appearance of a
real body, standing out in its solid state. An instrument has been lately invented by professor Wheatstone
for doing this: it is called the *stereoscope* (from two
Greek terms, "*stereos*" and "*skopeo,*" signifying "*solid*"
and "*to behold*"). By its means, pictures are seen as
solid things. Professor Wheatstone's stereoscope has

proved that man acquires his notions of the solidity of near bodies, by seeing pictures of different sides of them (so to speak) with his two eyes, and by combining these unlike pictures into one by an act of mind.

673. The eyeball is *lodged in a hollow socket* of bone, prepared for its reception beneath the front of the skull.

The eye looks out from its hollow socket, beneath the pent-house of the eyebrow; and it has lids which can be opened or closed in front of it, according to circumstances, and which instinctively shut down whenever anything approaches the eyeball that is likely to be injurious to it. It has also an apparatus provided for keeping its transparent portion constantly washed clean. The eyeball is lodged loosely in its socket, and has a service of muscles attached to it, by means of which it can be rolled up or down through one-third of a circle, and sideways through nearly half. This arrangement is made in order that the organ may be easily directed towards any object that its owner desires to examine.

674. The organs of special sense are placed in the face in order that they may be *near to the brain.*

The sensorium, which forms a part of the brain, is the seat of all sensation (559). Hence, it is convenient that the organs of special sense, which send their nerve tubules to it, should be placed near at hand, otherwise the large nerve branches would have had to run through a lengthened course, in which their very important and delicate structures would have been so much the more exposed to the risk of accident. The sensorium is situated within the skull, some little distance behind the forehead. The eyes look out in front immediately beneath the forehead. The ears are hung at each side of the head, so that they can readily catch the tremblings of sound, whichever way they come. The nose guards the opening to the chest, or breathing cavity, immediately beneath and between the eyes. Under the nose is the mouth, in which the tongue lies in wait for the savory morsels that are brought there on their way to the stomach. To all these organs the sensorium is connected by its myriad of nerve tubules. The face is thus the outer development of the creature's sensorial powers, so to speak. It is the great sentinel of the frame, constantly watching all that is going on around it, and constantly sending in the information that it acquires to the sensorium, in order that it may be made in the first place the subject of sensation, and in the next place of ideas (588). The general surface of the body, and in particular that of the hands, is auxiliary to the face in this service of observation.

675. Some kinds of information relating to external objects are acquired by the *combined operation of two or more of the senses* at once.

Every notion is originally acquired through observation made by one or other of the organs of sense (624). It has been shown that each special sense gives information concerning distinct and separate properties of bodies; but in addition to this, there are many circumstances that are learned only through two or more of the senses acting together. Thus, for instance, the eye and hand work in concert in ascertaining the real forms of objects that are scattered around. Children always attempt to handle what they are observing very intently. Most of man's notions regarding the mutual relations of surrounding objects are obtained by the combined and conjoined action of the different senses, so employed that the one aids or corrects the impressions of the rest.

CHAPTER XXII.

THE VOICE AND SPEECH.

676. Any thing that is capable of *making the air vibrate* or tremble, may originate sound.

To do so, however, the body must first tremble in its own substance. Thus a drinking glass becomes a sounding body when its rim is struck ; but this is because its own substance first vibrates, and then its vibrations are impressed on the surrounding air (638).

677. Musical instruments are bodies *purposely fashioned* for the convenient production of sound.

Hence the substances, of which musical instruments are made, are always of a highly elastic nature (639).

678. There are two different kinds of musical instruments, *the stringed and the wind.*

In the one of these, vibrations are first produced in a stretched string, and communicated from it to the air : the violin is of this kind. In the other, the vibrations are first produced in a confined column of air, and are then allowed to escape from it into the surrounding aerial substance. Flutes and organs are of this kind.

679. In stringed musical instruments the strings are *made elastic by stretching.*

That is to say, the ends of the strings are pulled so tightly opposite ways, that when they are disturbed, their middles vibrate backwards and forwards as a whole. It is not that the substance of the string trembles, but the string itself swings from side to side, and pushes the air before it.

680. The more tightly strings are stretched the *sharper is the sound* they produce when they vibrate.

This is because when they are very tightly stretched their elasticity is made so much greater, and they are forced by it to swing from side to side so much more rapidly (649). If the string of a violin or guitar is turned up tighter and tighter, and made to sound each time, it will be found that the sound it causes after each turn is sharper than the one produced before.

681. In wind instruments the elasticity of the column of air that vibrates first *results from its confinement in a closed tube.*

When air is pushed, it moves out of the way unless prevented by mechanical means from doing so. If it is so prevented, its substance yields instead, and gets compressed for the instant, but then leaps out again in consequence of its elasticity. When the mouth of a pipe full of air is blown upon, the air inside is made to tremble provided it cannot get out at the further end of the tube. This trembling is communicated to the external air, and becomes perceptible as sound. The tones of the organ are produced exactly in this way.

682. The sound produced, when a confined column of air is made to vibrate, is *sharp or low*, accordingly as the column is short or long.

The pipes of the organ which produce the sharpest tones are only a few inches long : those which produce the lowest tones are more than thirty feet long. This is because the larger the body of air that is primarily disturbed and made to tremble, the larger and slower are its vibrations; and the smaller the body of air the smaller and quicker its vibrations. In the flute, the confined, vibrating column is made longer or shorter at will, by opening or closing holes in different parts of the tube. The breadth of the tube and the size of its mouth also, in a degree, modify the character of the sound produced.

683. Some musical instruments are *half string and half wind.*

That is to say, stretched structures and confined columns of air are made to vibrate together, and the sound that is heard is a mixture of vibrations from the two sources. In these cases, however, membranes are used, instead of strings for the stretched parts, because then they can be made to form the openings of the tube, which could not be the case with strings. When a trumpeter sounds his instrument, he draws his lips tight over the end of the trumpet, and then blows past their edges, and sets at once those edges and the column of air contained within the tube vibrating. The result is the musical sound which floats to the ear. Sometimes tongues of elastic substance, attached at one end and projecting at the other over the opening, are used instead of the membranes. The tongues then tremble when blown upon, in consequence of the spring-like elasticity of their own substance. Their free ends play backwards and forwards without needing any stretching. In these cases the instrument is called a *reed* instrument. The stretched membranes and reedy tongues of musical instruments are only modifications of strings. The substance that produces the vibrations, plays backwards and forwards as a whole, and does not tremble internally. What are called the reed pipes of the organ are of this mixed nature. They have all vibrating tongues in their openings, and the sounds produced flow at once from these tongues, and from the columns of contained air. The clarionet and bassoon are other instances of mixed reed instruments, contrived for the production of musical tones.

684. The *organ of voice* is a musical instrument of a mixed kind.

The organ of voice, which is possessed by the most perfect animals, is a pipe with two stretched membranes placed at the sides of its opening. Air is blown through this pipe, and between the membranes, and by the blast both the edges of the membranes and the column of air itself are caused to tremble until vibrations of sound are spread around.

685. The pipe of the organ of voice is called *the windpipe.*

The windpipe is placed in the neck, and leads from the mouth to the chest. It is composed of rings of gristle and fibrous membrane, and serves for the admission of air into the lungs in the process of breathing. It is this air, thus employed for breathing, that is used in the production of vocal sounds. Impure air, that has been spoiled in breathing, must be got rid of; and nature has taken advantage of this necessity, and turned what would otherwise be altogether waste, to good account, in making it cause vibrations of sound, when needed. This is why the organ of voice is situated on the top of the breathing tube.

Fig. 88.

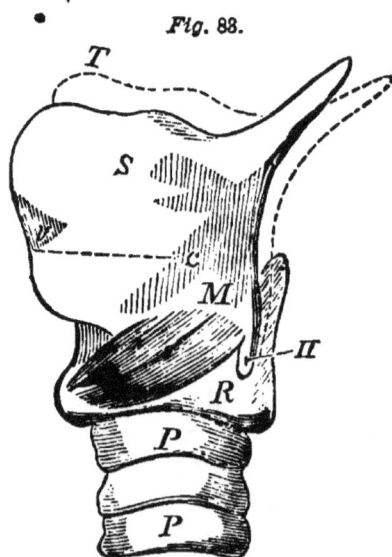

686. The windpipe is enlarged at the top into a *little box of gristle,* which is called the *Larynx.*

(The word *larynx* is derived from the Greek term " *larugx*," which is the name for the " *throat.*") This larynx is properly a frame made up of several distinct parts. In *Figure* 83 the form of the larynx is shown, as it appears from the outside. *P P* is the top of the windpipe, and *R S* are the gristly plates of the larynx.

687. The walls of the la**r**ynx are principally composed of *two separate pieces of gristle*, connected together by fibrous membrane. These two pieces of the larynx are represented in *Figure* 83. *R* is a ring-shaped piece of gristle, much deeper behind than before, and called the " ring-shaped cartilage,"on account of its form. *S* is a broad plate of cartilage folded round in front, but open behind where the back of the ring is placed. This is called the " shield-shaped cartilage," on account of its broad flattened form.

688. One of these pieces of gristle is capable of being *moved up and down* upon the other placed below and within it. The shield-shaped cartilage *S* (*Fig.* 83) is prolonged downwards into a sort of horn at *H*, and is fixed to the inner cartilage *R*, by a kind of pivot, at the end of the horn. The consequence is, that *S* can be turned on this pivot until it is raised into the position shown in the dotted lines. This movement. is effected by means of muscular cords that run from one piece of cartilage to the other in the direction necessary for the production of the result. Some of these muscular bands are shown at *M*.

689. The two pieces of gristle are connected together within, by *a pair of cord-like folds of membrane* stretched from the back of the one to the front of the other. The direction in which the membranes run is shown in *Figure* 83, by the dotted line *v c*, supposed to be seen

through the plate of gristle. The end *v* is attached to
the inside of the shield-shaped piece, the end *c* to the
back and top of the ring-shaped piece.

690. These membranes leave a *narrow
chink or slit between them*, which is the opening into the cavity of the larynx.

This is shown in *Figure* 84, which is a representation
of what is seen when the top of the larynx (*T* in *Figure*
83) is looked into. *S* is the
outer shield-shaped cartilage
open behind. *R* is the ring-
shaped cartilage contained
within. *v. c* are the two membranes stretched from the inside front of *S* to the back and
top of *R*, and between them
is the narrow chink which
forms the opening into the interior cavity of the larynx,
and through it into the wind-
pipe. All the air that passes
in breathing goes out and in
through this chink. which is called the "*glottis*," (a word
derived from the Greek term "*glotta*," which signifies a
"*tongue*," or "*speech*.")

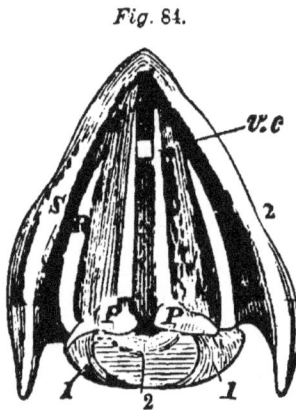

691. The membranes, stretched on each
side of the opening of the larynx serve as
reeds to the organ of voice, being made . to
vibrate in the production of its sounds.

These reed-like membranes are called the "vocal
cords." They are made elastic in the same way as the
strings of a violin, that is, by being stretched tightly
from *v. c* to *P P* (*Figure* 84). When they are stretched
tightly, and air is blown with force through the narrow
slit between them, their edges are made to tremble,
just as the edges of the tightened lips do when the trumpet or horn is blown. The air contained within the cav-

Fig. 84.

ity of the larynx and wind-pipe gets compressed at the same time, owing to the narrowness of the outlet through which it is forced, and trembles also. By the combined influence of the trembling air and membranes, the sound known as voice is formed.

692. The membranes, that form the sides of the opening of the larynx, can be *stretched tighter*, or *made looser*, by the movements of the shield-shaped cartilage.

The attachment of the vocal cords (*v c*), *Figure* 83, is made to the pieces *S* and *R* in such a way, that when the top of *S* is raised, by being turned upon its pivot *H*, into the position marked by the dotted lines, the string-like membranes are loosened. When it is depressed away from *T* the membranes are tightened. When the membranes are tightened, they vibrate more quickly, and give out sharper sounds: when they are loosened, they vibrate more slowly, and give out lower sounds. All the musical notes of the human voice are produced in this way, by altering the degree in which the vibrating cords of the organ are stretched. The varying force with which the air is blown between them, and slight modifications in the position and length of the wind-pipe, assist in an inferior degree in the production of this result.

693. When the vocal membranes are not in use in the production of sound, *the opening between their edges is widened*, and they are themselves loosened until they lose their elasticity for the time.

If this were not the case, the passage of the air during the act of breathing would be always attended with noise. In *Figure* 85 the state of things is shown which is present when the opening of the larynx is only used for breathing, and not for singing or speaking. The backward ends of the vocal cords are attached to the ring-shaped cartilage by means of two little pyramids of gristle, $(P\ P,)$ which lie, as sketched in *Figure* 84, when the vocal cords are in use ; but are bent, as shown in *Figure* 85, when the membranes are not employed. When the opening of the larynx is only used for breathing, it is wide and of a four-sided lozenge-like figure. When it is employed in the production of vocal sounds, it is a narrow slit. It is kept fit for quiet breathing by the contraction of the little muscular bands, marked 1, 1, *Figure* 84. And it is prepared for issuing sounds by the contraction of the muscular bands marked 2, 2. As soon as the voice is needed, these muscles are set to work, the vocal cords are brought near together, their edges are made tight, and the air passing between them then sets them trembling and vibrating. In this way it is that the audible tones of the voice are formed.

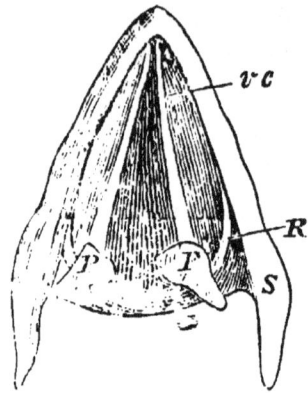

694. The small organ of voice is the *most wonderful and perfect* of all known musical instruments.

This is proved by the performance of clever singers. In flute sounds the tones can be sustained for a length of time, but their intensity cannot be varied. No force or expression can be communicated to them. The intensity of stringed sounds can be varied, but they cannot be sustained. The tones of the human voice can be both sustained and rendered more intense, by blowing the air through the chink of the organ with greater force and by tightening up the cords at the same time. In the organ a great range and variety of power is attained, but this is done by the employment of hundreds of flute and reed pipes, each of which can only be used for the production of one tone. In the little organ that man carries in his throat, all this is more perfectly accomplished by a single pipe, and by one pair of membranous strings. The different musical notes are called out by altering the tightness of the membranes, and by modifying the force with which the air is blown between them. Diversity of pitch, in different voices, depends on the different length of the vocal cords. Variety of tone is caused by the varying quality of the material of which the cartilages and membranes are made, and by the difference of size in the larynx.

695. The vocal sounds are capable of being *considerably modified* in various ways, *as they pass through the mouth.*

If the tongue, cheeks, and lips, are placed in different positions while a vocal tone is sustained, it will be found that the tone varies with each instant, so that the ear can detect the difference.

696. The distinct modifications of the vocal tones, effected by altered positions of the mouth, are called *articulate sounds.*

The mouth is a cavity formed of exceedingly loose and movable walls, and having an opening that can take upon itself a very great difference of size and shape. This mouth forms the outlet of the organ of voice. The vocal sounds are produced by tremblings of the air and membranes of the larynx; but they are then moulded in various ways as they pass through the appended cavity, and issue from it. Each altered or moulded tone is called an articulate sound. It is termed in this case " articulate," because it is capable of being joined to or hung upon its fellow sounds, which the pure vocal tones cannot be, so far as they are exclusively concerned. (The word "*articulate*" is taken from "*articulo,*" the Latin for " *to joint.*")

697. Articulate sounds, jointed together in various ways, constitute *language or speech.*

M*

" *Language*" is derived from " *lingua*," the Latin term for " *tongue ;*" this organ being one of the parts principally concerned in forming the sounds of which language is composed. " *Speech*" is a modification of the old Anglo-Saxon term " *spæcan.*"

698. The articulate sounds of language are *signs* that are taken to stand for certain *agreed-upon meanings.*

When messages were sent by the old plans of telegraphic communication, an arm of wood, that could be seen from a great distance, was placed in a given observable position, which it had been before determined should mean some particular thing. There was, in this case, no absolute relation connecting the visible appearance with the notion conveyed. The relation was merely a matter of artificial arrangement, agreed upon for the sake of convenience. Articulate sounds stand exactly in the same position. They have no absolute relation with the meaning they are understood to express. They are merely artificial signs, which are taken by mutual agreement to represent certain notions and ideas, with which they have been associated. The sound and the object or meaning have been connected together in the mind, and it has been understood that for the future that particular sound shall always call up the same meaning, whenever it occurs. Articulate sounds are *audible signs*, by which intelligent creatures telegraph their messages to each other across intervening tracks of air. Advantage is taken of the peculiar construction of the organ of hearing, and of its relations to the elastic properties of the all-pervading air, to " sound messages afar," instead of to " write them afar" (*tele-grapho*). And for this purpose an extensive code of audible signals is contrived. Speech is the great means of communication whereby intelligent man expresses the ideas that are present in his mind, so that they may be entertained also by, or may influence, his neighbor.

699. Articulate sounds *are grouped together to form words.*

The great value of articulate sounds depends on the readiness with which they may be "articulated" or jointed together. There are 27 distinct and distinguishable simple sounds in the English and in other languages, that are formed by altered positions of the mouth as the vocal tones issue through it; but these 27 simple sounds are clearly insufficient for the purposes of speech. They would only answer if there were but 27 notions to be conveyed by them. The notions that are conveyed by language are many thousands, instead of 27. Hence, the 27 elementary sounds are changed about amongst each other in all conceivable ways, and thus many thousand compound sounds or words are made out of them, each of which words then becomes the audible signal to which an agreed-upon meaning is attached. The changes are rung upon the audible sounds, just as changes are rung upon peals of bells. Any one who knows the immense variety into which the tones of a full peal of bells can be arranged, will at once understand how infinite must be the power of 27 distinct sounds in word-making. Twelve bells can be rung 479 millions of ways, before they need to be chimed in the same order for the second time. It has been calculated that it would take 91 years to ring through such a series of changes! and that it would require 117 thousand billions of years to ring through the changes of 24 bells! This then gives an idea of the number of changes that could be rung on 27 articulate sounds. The number of words that could be formed by them is almost beyond human powers of computation. Hence, only a very few of the elementary sounds are commonly taken at once to be articulated into a word: out of these few many thousands of convenient words are readily formed.

CHAPTER XXIII.

RESPIRATION.

700. The impure venous blood is sent by the right side of the heart *to the lungs* (420).

The word "*lungs*" is taken from the old Anglo-Saxon "*lungena*," itself derived from "*langen*" "*to draw.*"

701. The lungs are placed with the heart in the *cavity of the chest* (388).

The chest, it will be remembered, is the cavity of the body which is situated above the arched ceiling of the abdomen (*D, Figure* 26). This cavity is almost entirely filled by the heart, lungs, and the great vessels that are in communication with these organs. The reason for the heart and lungs being associated together in one cavity is an economical one. One half of the blood pumped out from the heart at each pulsation, goes to the lungs alone, and the only object for which it is sent there is free communication with air. It is, consequently, an obvious convenience of arrangement that the lungs should immediately surround the heart; otherwise the large quantity of blood would have needed to be forced through a long distance, without any object being answered thereby. The chest is entirely devoted to the circulation and purification of the blood, as the abdomen is to the processes of digestion (299). *Figure* 86 shows the position of the several parts that are contained

Fig. 86.

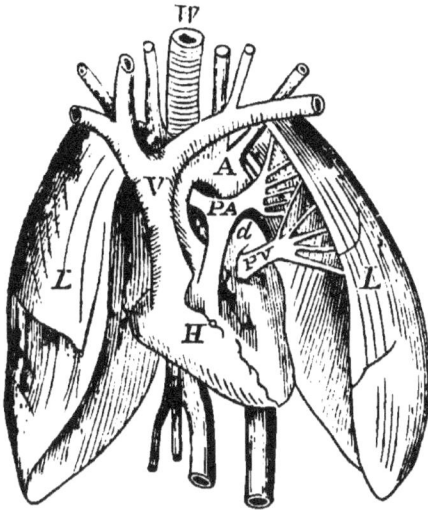

within the chest. L L are the lungs of the right and left side. H is the heart placed between them. V is the great trunk vein that brings the im pure venous blood from all parts of the frame to the right side of the heart. P A is the pulmonary artery that conveys the im- pure venous blood to the lungs for purification (R, Figure 37); it divides at d to run right and left to the lung of either side. P V is the pulmonary vein that carries the purified blood back to the left side of the heart, and A is the great artery that issues the crimson stream to every part of the body.

702. The chest is composed of a frame-work of bones, lined by serous membrane (453), and covered by muscles and skin.

Fig. 87.

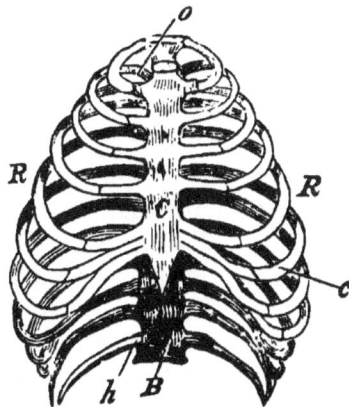

The chest is not a bag with soft and flexible walls, like the abdomen (299). Its membranes are stretched over a rigid frame; hence it derives its name. ("Chest" is taken from the Latin "cista," which signifies a "basket," or "box.") Fig-

ure 87 represents the frame-work of bones upon which the walls of the chest are moulded. *B* is the backbone, or spinal column, which is the grand support of the cavity behind. *C* is a flat plate in front known as the breast-bone. *R R* are curved pieces, 12 on each side, which run from the backbone to the breastbone, leaving a great hollow space between their concavities : these are called the ribs.

703. The chest is capable of being *expanded and contracted* by the movements of its walls.

Although circumscribed by a rigid frame, the cavity of the chest can be enlarged and diminished, because that frame is formed of several pieces which are so hinged together as to allow a certain degree of move-ment between each other. The ribs are all jointed to the backbone at *h* (*Figure* 87), and they are united to the breastbone at *c* by pieces of gristle, which are elas-tic and yield like india-rubber. Muscular cords are so attached to the arches of the ribs that they can be made to play all together up and down upon the backbone by means of their joints, and the ribs themselves are set obliquely downwards upon the backbone, so that when they are raised or depressed, the space contained within their concavities of necessity becomes successively larger and smaller. In addition to this, the floor of the chest is arched up, as shown at *D, Figure* 26. This arched-up floor is formed of interwoven muscular fibres, which can alternately draw down and flatten the arch by their con-traction, and bend it up by their relaxation. By this double means of movable walls and a movable floor, the chest is formed so that its size can be varied con-siderably. The arched muscular floor of the chest is really a partition between its cavity and that of the ab-domen (*D, Figure* 26) ; hence it is called *the diaphragm*. (The name is taken from two Greek words, "*dia*" and "*phragma*," which signify "*between*," and "*a parti-tion*.")

704. The wind-pipe, which is connected with the back of the mouth (685), *runs down* through the neck *into the cavity of the chest.*

In *Figure* 86, *W* is the top of the wind-pipe. This tube finds its way down in the front of the neck, before the swallow, and passes through the opening *o* (*Figure* 87), left between the top pair of ribs. In this way it enters the chest.

705. In the chest, the wind-pipe *branches out* into numerous small tubes, which form the *air tubes* of the lungs.

Just above the heart the wind-pipe forks into two main branches, one for the lung of each side. Each of these branches, having entered the fleshy substance of the lung, divides more and more, after the fashion of the branches of a tree, until at last it forms a great abundance of small twigs, so minute that about one or two hundred of them would lie in the extent of an inch. These branches and twigs constitute what are called the *air tubes* of the lung. They run through its substance everywhere, and form, indeed, a sort of internal frame or skeleton, on which all the softer parts of the organ are arranged. The wind-pipe and its subdivisions, the air tubes, are all composed of fibrous membrane (437), strengthened and made rigid by means of little rings of gristle, which serve also to keep their interiors constantly open for the ready reception of air.

706. Each air tube of the lung terminates in *a dilated bulb-shaped cavity or cell.*

Fig. 88.

The dilated cavities to which the air tubes lead, are called the *air cells* of the lungs. The manner in which the air tubes terminate in bulbous cells is shown in *Figure* 88. These cells are formed of very delicate films of membrane. There are no rings of gristle in them, as in the air tubes.

They are like little soft flexible bags, hung upon the ends of the rigid tubes. Every portion of the fleshy substance of the lungs is honey-combed by these cavities. It has been calculated that the lungs of a full-grown man contain no less than 600 millions of them; and that, small as they are, if the membranes of which they are composed were all spread out flat as one continuous sheet, this would cover 21 thousand square inches of surface. It would be some 30 times more extensive than the entire tract of the skin. These cavities are called air cells, because they are always more or less distended with air.

707. The blood-vessels, that carry impure venous blood from the right side of the heart to be purified in the lungs (420), are distributed as *capillary vessels on the filmy walls of the air cells.*

Figure 89 represents an air cell of the lung, terminating an air tube (706). It also shows how a net-work of very fine blood-vessels runs all over the filmy wall of the cell. The great vessel, that issues from the right side of the heart to carry the venous blood to the lungs (*P A, Figure* 86), divides like the wind-pipe into two main branches, and each of these then run into the lung of its own side, and there ramifies in the midst of its fleshy substance more and more, until at length it forms a sort of net-work of fine branches, containing the air cells in its meshes. From the meshes of these branches numerous capillaries are sent off, to connect themselves closely with the thin film of the air cell, as represented in the figure. These capillaries lead to veins exactly as the capillaries of the general system do (400), and these veins are gathered together to form the main trunks of the vessel that carries the purified blood from the lungs to the left side of the heart (*P V, Figure* 86). As the right chambers of the heart dilate and contract, venous blood pumped out

Fig. 89.

by them is made to flow in a steady continuous stream through the capillary vessels that are distributed to the air cells of the lungs, exactly as the arterial blood does through the capillaries of the general frame (391, 396), when pumped out of the left side of the heart. The substance of the lungs consists of air tubes and cells, arteries, veins, and capillary vessels, connected together by fibrous membrane, and bound up in an external covering of the same.

708. So long as the blood is within the capillary vessels of the lungs, it is in *close communication with the air* contained within the air cells.

The capillary vessels, sketched in *Figure* 89, are filled with blood, and the little bag or air cell on which they lie is distended with air; consequently the blood in the tubes, and the air in the cell, are only separated from each other by the conjoined walls of the two.

709. The blood and the air, that are in close communication with each other in the lungs, *mutually influence each the other.*

The conjoined films of the blood vessels and air cells, which separate the blood from the air in the lungs, are together of extreme delicacy, and although sufficient to retain their contents within their cavities, as a bladder holds syrup tied up in it (162), when not influenced by any power of greater force than gravity, they are not sufficient to prevent intermixture under the operation of osmose (162). In the lungs, indeed, there are present the exact conditions that are essential to set up the movements of osmose. There is a fine double film of membrane with different kinds of fluid (air and blood) on its opposite sides (164). The consequence is, that portions of the air pass into the vessels to mingle with the blood, and portions of the constituents of the blood pass out of the vessels to mix with the air in the cells.

710. A *portion of the oxygen* of the air is

20

sucked by the action of osmose into the blood contained within the capillary vessels.

The air that is taken into the chest in breathing contains 21 parts of oxygen in every hundred parts (eight grains to every thirty-six) [57]; but the air that is thrown out from the chest in expiration, contains only thirteen parts of oxygen in every hundred parts. The other eight parts have been removed from the air while lying in the interior of the air cells, to be mixed with the blood. Nitrogen also is taken into the blood, but, being very inert in its chemical properties (50), it passes through the circulation unchanged, and is returned to the air again. As much nitrogen is given back from the blood to the air (by exosmose, 167), as is taken from it by osmose, and, consequently, air that is expired contains about the same proportion of this element as that which is inspired. The quantity of nitrogen in the air is not altered by its being breathed; the quantity of oxygen is diminished by nearly one half.

711. The oxygen that is absorbed from the air into the blood in the lungs, is chiefly *attached to the films of the red corpuscles* of that liquid (335, 411).

The films of the corpuscles are made thicker and whiter by the union of oxygen with their substance.

712. Dark venous blood *is changed into bright arterial blood* in the capillaries of the lungs.

This change of color is effected by the oxygen making the walls of the corpuscles brighter and more reflective of light when it is united to their substance (412).

713. The oxygen communicated to the blood corpuscles in the lungs, *is conveyed by them to all parts of the frame.*

The purified and arterialized blood is sent back from

the lungs to the left side of the heart, and is by it pumped out to all the structures of the body, to supply them with nourishment, and to support their vital powers (424).

714. The oxygen, conveyed to the various structures of the body by the arterial blood, *consumes their substance* by the , exertion of its strong chemical energies (35).

That is to say, it unites itself chemically with the elements of their complex molecules, and so reduces them into simpler compounds. It eats away the materials of muscles, of nerves, and of brain, and converts them into water, carbonic acid, and other oxides (254); out of their destruction it produces animal force—that is, voluntary motion, sensation, and perception, and other like activities (258). It is from the lungs, then, that the arterial blood acquires that redundant oxygen, which it carries with it, wherever it goes, to be the great *stimulus or good* to animal life. It is in the narrow channels of the capillary circulation that the arterial blood gives up its redundant oxygen to attach itself to the various structures of the organic frame (399). Hence it is, that arterial blood is changed into venous blood in these channels (409).

715. When oxygen corrodes, or consumes the substance of complex organs, *heat is set free* and rendered sensible in consequence of the molecular change.

When oxygen is united with compounds of carbon and hydrogen—such as coals, tallow, and oil—quickly and at high temperatures, the effect is attended by a production of light and heat (47). The light and heat so set free have been lying hid in the substance of the combustible bodies for long periods of time, during which they have been employed in holding the substance in its existing state, instead of in producing light or heat that could be seen or felt. Both the light and heat were, however, originally derived from the sunbeams. When the plants grew, that produced the wood (since changed to coal) and the oil, or that fed the animals which have supplied their fat for tallow, they did so in a warm bright atmosphere, from which they took light and heat, at the same time that they imbibed their organic elements (122), and the whole were fixed and mingled together by the plants, in the form of the complex compounds. When a complex compound is ultimately dissolved into its several elements, these are all separated from each other, and made appreciable. Light and heat, consequently, are set free among them. They are liberated from their secret and inappreciable work, and are allowed to show themselves and to make themselves felt. The same thing happens in a degree in the living animal frame. The gluten, fibrin, albumen, hæmatin, and other complex principles, have all been originally made by the operations of vegetable life (245) under the stimulus of sunshine, and some of the warmth of that sunshine has been stored away in them insensibly. When, however, those complex principles are resolved by the corrosive powers of oxygen into simpler compounds, their several elements (the warmth among them) are set free, and made distinctly appreciable.

716. The heat that is set free, when the complex structures of the living frame are consumed under the influence of corrosive oxygen, is *employed in warming the blood.*

The union of the oxygen with the materials of the organic structures·takes place in the capillary vessels, or just outside of their walls (399); consequently the heat that is set free is at once thrown into the blood moving through those vessels, and warms it to a certain extent. But the capillaries of the blood-vessels extend to all parts of the frame (395). There is blood everywhere; therefore the entire body is at nearly the same heat. The oxygen, that is conveyed to the general capillaries by the arterial blood, fans a smouldering flame, which is never altogether extinguished during life. Animals carry about in their insides little furnaces, which keep their bodies warmer than the surrounding air.

717. The human body is kept at a temperature *two-thirds of the way from ice towards boiling water*, by means of the heat set free in the blood.

The difference between the temperature of freezing and boiling water is adopted as a standard for a heat-measuring scale, and is reckoned as 180 degrees (79). The mean temperature of the air in England is about sixteen of these degrees warmer than frost. But the human body that exists in this air is kept steadily sixty-eight of the same degrees warmer than frost. As much heat is produced in a single year in the blood of a full-grown active man through the burning of the fuel and structures of his frame, as would be sufficient to change eleven tons of ice into boiling water. This proves just enough, under existing arrangements, to keep his body at a steady temperature of 100 degrees of the heat scale. (100 degrees, because frost begins to reckon as 32 degrees).

718. The burning of the combustible materials of the animal body is *a very slow and gradual process* as compared with the burning of ordinary fuel.

The internal furnace, that keeps the animal body warm, is a very slow one. The fuel that is consumed is of a nature to admit of a very gentle smouldering, rather than of a rapid consumption. This is the only real difference between the burning of the living body, and the burning of fuel in a common fire. Coals and oil burn so fast that they produce intense degrees of heat as well as light, in consequence of the rapid change. But living flesh and blood burn so gradually that no light, and only blood-heat results from the combustion.

719. A portion *of the blood itself is burned* in the production of the warmth of the body, as well as the more finished structures.

The portion of the blood that is burned in the production of heat, is its oily constituents (371, 372). A steady temperature of 100 degrees is essential to the healthy performance of the various operations of the frame. A part of the heat necessary to produce this is afforded by the burning of the organized structures of the body. But with the ordinary habits of the human creature, enough heat would not be furnished from this source; consequently an additional supply of fuel is thrown into the blood, which has nothing else to do but to be burned in the production of heat. This is furnished by the oily and farinaceous ingredients of his food (303, 305). Carnivorous creatures, that undergo violent and long-continued exertion, get very much more of their heat out of the destruction of their own frames, and less out of the simple burning of the fuel of the blood than man and herbivorous animals.

720. When the materials of the organized structures and the fuel of the blood are burned in the living body, *carbonic acid is formed* as a result of the burning.

When combustible substances containing carbon are burned in the presence of oxygen, portions of that element invariably unite with successive quantities of the carbon, and carry them off in the condition of carbonic acid gas (108). All the combustible principles of the animal body do contain carbon in very large proportions; consequently carbonic acid is thrown into the blood when they are burned. Whenever arterial blood is changed into venous blood in the capillary vessels of the general system (409), it loses its redundant oxygen (410), but it acquires an equal amount of redundant carbonic acid instead (414). Bright arterial blood is overcharged with oxygen. Dark venous blood is deficient in oxygen, and overcharged with carbonic acid, for carbonic acid is one part of the waste materials, formed wherever oxygen unites with the tissues of the body, and therefore needing to be carried away.

721. The redundant carbonic acid of the venous blood is *exhaled from the lungs* in breathing, and is so got rid of out of the system.

The air that is expired, or thrown out from the lungs at each breath, has less oxygen in it than that which was taken in (710). But it has much more carbonic acid in it than it had before it was inhaled. Pure air has only one part of carbonic acid in every two thousand parts of its substance (100); but air, that is just thrown out from the lungs in breathing, has as much as one hundred parts of carbonic acid to every two thousand parts. The carbonic acid, produced by the burning of the fuel within the body, is poured out from the lungs by the mouth, exactly as that produced by the burning of the coals of a fire is out of the chimney. One pair of lungs pours out into the air in this way nineteen cubic feet of carbonic acid every twenty-four hours; that is, as much as would nearly fill four three-bushel sacks. This redundant carbonic acid is exhaled from the blood contained in the capillary vessels of the lungs, into the cavities of the air cells, under the operation of exosmose (167). It, then, is one of the constituents of the venous blood that escapes from that liquid to mingle with the air (422).

722. The removal of carbonic acid from venous blood is one of the means by which that liquid *is purified in the lungs* (420).

In the nineteen cubic feet of carbonic acid thrown out from the lungs every twenty-four hours there is as much as eight ounces of solid carbon! By means of respiration the blood is thus relieved of eight ounces of this dense substance every day. It will at once be understood how important a process of purification this must be, if the question be asked what would soon become of the circulating fluid if the removal were stopped, and eight ounces of solid carbon were added to the liquid every successive day. In three or four days alone, it would become too thick to flow through the delicate channels of the capillary circulation.

723. *Watery vapor* is also thrown out from the blood in the capillary vessels of the lungs.

The air, that is driven out from the chest at each breathing, is loaded with moisture as well as with redundant carbonic acid. It always holds as much steam suspended in it as air at the temperature of 100 degrees is able to retain (84). In cold days this steam is seen to be condensed into visible mist as it issues from the mouth. The vapor escapes through the membranes of the capillary vessels and air cells into the cavities of the latter under the simple operation of exosmose (167). From sixteen to twenty ounces of water are exhaled from the blood through the air cells of the lungs every twenty-four hours.

724. The vapor that is exhaled from the lungs is *tainted by impurities*.

As carbonic acid is soluble in water (99), this exhaled vapor contains as much of the redundant carbonic acid as it can dissolve; but it also contains putrefying organic matter that has been mingled with it in the blood. If the vapor of the breath be condensed into a liquid, and the liquid be kept for a little time in a warm place, it soon gives off a putrid smell. The disagreeable odor of close air, that has been breathed over and over again, is due to the accumulation of this exhalation. One thousand parts of expired vapor, contain generally three parts of this peculiar impurity.

725. The air cells and tubes of the human lungs contain *about two hundred and twenty cubic inches* of air within their cavities.

That is, about the bulk of three quart measures. A

full-sized and fully-expanded chest of a man has within it three quarts of air. The rest of its cavity is filled with blood, and with the membranes of its tubes and cells.

726. The air contained within the cavities of the lungs *is continually being changed* by the movements of respiration.

It is necessary that it should be so changed continually, for so long as it remains in contact with the walls of the air cells and capillary blood-vessels, it is being more and more contaminated by the loss of its oxygen (710), and by the addition to it of carbonic acid (720) and tainted vapor (724). The change of the air is effected by the movements of the walls of the chest, carried on through the instrumentality of muscular contraction (703). When the walls of the chest are raised and its floor is flattened, the interior cavity is made a little larger, and air flows in to fill the increased space; when the walls are drawn down and the floor is arched up, the interior cavity is diminished, and air is driven out. These alternate movements are made about fifteen times every minute, and continue both in sleep and during wakefulness. They are carried on through the influence of a series of nerves that come from the upper prolonged portion of the spinal cord (557), which it will be remembered is the seat of unconscious actions, and they are excited by the impressions impure air produces when resting in the air cells. This action of changing the air contained within the lungs is called "*respiration*." (The Latin word "*respiro*" means to "*breathe*." "*Breath*" is a modification of an old Anglo-Saxon word, "*bræthe*.")

727. About *a tenth part* of the air contained within the chest is changed at each ordinary breath.

The entire volume of three quarts of air is not pumped in and out at each breath. The chest is kept tolerably

N

distended with air at all times, and a small portion only of this is changed at each breathing. In quiet, unforced breathing, about 20 cubic inches (nearly two-thirds of a pint) is changed at each contraction and expansion. The fresh air admitted each time mingles at once with the entire quantity contained in the chest, under the law which makes unlike gases mix together (58), so that in this way it is kept in a state of tolerable purity.

728. *Nearly one half* of the air contained within the chest can be changed at once by a forced breath.

Although the breathing is generally an involuntary and unconscious operation, it can be deepened by an act of will. The ordinary shallow breathings are not sufficient to purify the blood, even in a state of repose, and hence people generally make a deep sigh every now and then to assist the operation.

729. About *three hundred cubic inches of air are spoiled every minute* by the respiration of one pair of full-sized human lungs.

A full-grown man spoils rather more than a gallon of air every minute by his respiration. This amounts to 250 cubic feet, or nearly fifty three-bushel sacks full in twenty-four hours. If a man could be made to breathe for twenty-four hours in a room *seven feet square* all ways, which was so completely closed up everywhere that no change of air could take place, the entire quantity of air would, at the end of the twenty-four hours, have been changed into the condition of air just expelled from the lungs. That is, it would have in it 100 parts of carbonic acid, and 300 parts of oxygen to the two thousand, instead of one part of carbonic acid and 420 parts of oxygen to the two thousand, which is the case with pure air (710, 721). It would also be laden with the peculiar putrefying exhalation of the breath (724).

730. Air that has been spoiled by being

breathed is *no longer suited for the purposes of respiration.*

Human beings cannot continue to breathe air that contains only 300 parts of oxygen, and that has 100 parts of carbonic acid to the two thousand, besides being loaded with putrefying exhalations. On this account, a man could not live for twenty-four hours shut up completely in a room seven feet square all ways. He would die long before that time, poisoned by the products of his own breath.

731. Air that has been spoiled by breathing *is injurious in a three-fold way,* if the attempt be made to breathe it over again.

In the first place, it does not contain so much oxygen as it ought; hence, when drawn into the air cells of the lungs, it does not supply so much of this energetic agent as the blood requires to keep it arterialized, and to enable the system to be relieved of its useless load of carbon (722). In the second place, it carries back into the air cells carbonic acid, which has been just expelled from the system as an injurious principle. And in the third place, it carries back the putrefying exhalation of the breath, which has just been driven out for the same reason. When these injurious compounds are taken into the air cells, they get sucked into the blood, through the walls of the capillaries, by the influence of osmose, exactly as fresh oxygen is (710).

732. Air that contains *one part of carbonic acid to every hundred cannot be breathed* for long periods without producing discomfort and ultimate mischief.

Not only is air which has been already breathed, and which contains five per cent. of carbonic acid (that is, 100 parts to the two thousand) poisonous to animated creatures. Even when mingled with four times its bulk of pure air, and therefore containing only one per cent. of carbonic acid, it is still very injurious. If a man were completely shut up for twenty-four hours, in a room *twenty feet square* all ways, but in which there was no ventilation, the air surrounding him would be unfit to be breathed at the end of that time. It would contain too little oxygen, and too much carbonic acid and animal exhalation to be able to play its proper part in the continuance of the vital actions of his frame.

733. Absolute *purity of air* is essential to the healthy existence of animal life.

If health is to be maintained uninterruptedly, the air that is breathed must be kept at nature's own standard of excellence: that is to say, it must have in it twenty-one per cent. of oxygen, and only half a part in the thousand of carbonic acid; and it must be quite free from all disorganizing taint. This natural standard of excellence is only found in the open air of the country. So many people live together in confined spaces in towns, that the air there breathed nearly always contains too little oxygen, and too much carbonic acid and impure exhalation. The result of this is that, as a general rule, people live many years longer when they dwell in the country than when they dwell in towns. It has been found by observation and experience, that a man who spends all his days in the open country, may fairly expect to *live seventeen years longer* than one who dwells entirely in a close city.

734. In the open country, the air is kept pure *by the influence of the wind, rain, and vegetable life.*

The wind mingles every vapor that is thrown into the atmosphere with the general bulk of its substance, and rain washes all soluble matters out of it as it falls, and deposits them in the soil. Growing plants feed upon the carbonic acid, fix its carbon in their structures, and return the oxygen pure (246). This is nature's plan for maintaining the great aerial reservoir in the precise condition in which animal life requires it should be preserved.

735. In the closed spaces of dwellings, and in towns, *artificial arrangements need to be made* for securing the constant purity of the air.

In the apartments in which he dwells, and in close towns, man finds it necessary to do what nature is constantly effecting in the close spaces of his own chest (726); that is, to *change the air* continually, by forcing out portions of the old, and bringing in an equal quantity of fresh to take its place. This changing of the air in the close spaces of human dwellings, is called "ventilation." (The word is derived from the Latin term "*ventilo*," which means "*to blow*," or "*fan*."

736. In warm, summer weather, the ventilation of dwellings takes place *through open doors and windows*.

During the warm season, the heat makes people open their windows and doors, so that then a free play of air occurs, and the atmosphere in the interior of rooms is kept nearly as much under the influence of moving currents, and nearly as pure as that outside.

737. In winter ventilation is principally effected *through the instrumentality of fires*.

A portion of the air in the room is consumed in the support of the burning of fuel, and the products are carried up the chimney. Fresh air enters at the same time through the crevices of the doors and windows to the same extent. It is not, however, as is generally conceived, that the hot air mounts in the chimney, and that the fresh air *in consequence* comes in through the doors and windows to *take its place*. The fact is, that the cold air comes in in spite of the resistance of the hot air, and drives it out up the chimney before it. Cold air is heavier, bulk for bulk, than hot air. Now, as air transmits pressure equally in all directions (65), the cold air outside a room presses through the crevices of the doors and windows upon the warm air contained within. If there be no other outlet from the room, the warm air inside acts like an elastic cushion, and resists the pressure of the outer air piled against the crevices, and there is no motion; but if there be an open chimney and a fire burning in the grate, the column of air over it gets very hot, expands and grows light. Then its weight is not sufficient to support the weight of the cold air pressing on the outside. The chimney is a weak place, so to speak, in the resistance. The lighter air, consequently, is squeezed in a stream up through it, by the heavier cold air pushing its way in through the crevices. If, in *Figure* 1, *b* and *c* were pressed upon with twice the force that *a* was, *a* would have to give way and rise, while *b* and *c* rushed in beneath it. In this figure, under such circumstances, *a* represents the column of light warm air in the chimney, and *b* and *c* the columns of cold heavy air pressing in through the external crevices. The ventilation which a fire effects in a room, depends upon the air, which the room contains, being unequally pressed on from without down the chimney and at the doors and windows, in consequence of the superincum·bent weight being made less in the chimney through the expansive power of heat. The air of the room flows in the direction in which the pressure is the least, and so produces the ventilation.

738. The ventilation *in closed rooms with-out fires must always of necessity be very imperfect* and insufficient for the purposes of health.

This is the case even although the same external crevices, and the same chimney openings exist, because then there is not the same difference of pressure in op-posite directions, to make the air flow up the chimney. If persons are living in the room, the warmth of their bodies raises the temperature of the internal air a little, and so a slight current is produced up the chimney; but this is not strong enough to effect sufficient ventilation to keep the air at its natural standard of purity. The

consequence is, that it soon gets laden with carbonic
acid and the impure exhalations of the breath, and becomes deficient in oxygen.   In such rooms some contrivance must be adopted for actually pumping out the
contaminated air, or for forcing in pure air if the in-
mates are to be kept in health.

739. In inhabited rooms, where no mechanical plan is adopted for free ventilation,
never less than 800 *cubic feet of air* must be
allowed for every pair of lungs.

And even this would need to be changed at least once
every twenty-four hours.   It has been seen that one pair
of lungs changes about 250 cubic feet of pure air, into
air containing five per cent. of carbonic acid, in twenty-
four hours; but such air cannot be long breathed.
Consequently, very much more than 250 cubic feet must
be allowed where rapid change does not take place.
The smallest allowance that can be made with safety is
800 cubic feet.   This would be contained in a square
room a little more than *nine feet* all ways.

740. In densely peopled towns, ventilation
can never be neglected, *without incurring
injury to health and destruction to life.*

In a general way, the ventilation of dwelling-houses
and factories in towns is very imperfect, and hundreds
and thousands of people die before their time in consequence.   The numbers of people who are killed by impure air in England is almost too great to be believed.
In the prisons of India there are generally about forty
thousand prisoners confined.   Never more than 300
cubic feet of air is allowed for each of these prisoners
daily, and often not more than 70 cubic feet, on account
of the crowded state of the establishments.   The result
is, that one in every four dies each year.   What happens in an extreme degree in these prisons is taking
place in an inferior degree in nearly all the free dwellings of our towns.   Wherever living creatures are

crowded thickly together, disease and death are sure to
be called into activity, unless powerful means are devised for supplying pure air artificially.

741. No system of ventilation is perfect
in which provision is not made for the *outward flow of the contaminated air*, as well as
for the *inward flow of the fresh air*.

It has been seen that this provision is made when
fires are burned in open grates (737). All plans that
are adopted for artificially ventilating dwelling-houses,
must proceed upon the principle that the use of the open
fire has introduced without design—namely, the production of a current or flow of air through the rooms. Two
opposite openings must be formed, and air must either
be forced into one, or it must be pumped out through
the other of these. In making arrangements to carry
out any system of artificial ventilation, it should always
be borne in mind that every pair of human lungs spoils
fifty sacks-full of air every twenty-four hours. A square
hole half an inch across both ways would just allow this
quantity to pass through it in twenty-four hours, if moving at the rate of twenty inches each second, the velocity
of an almost imperceptible breeze.

742. The opening, designed for the outward passage of the impure air in ventilation, should *always be made as near to the
ceiling of an apartment* as possible.

Carbonic acid is heavy and flows downwards through
air readily like water. The other impure exhalations
(724) of the animal breath are light and volatile, and
more readily rise upwards through air. On this account
the ventilation of rooms, in which ordinary fires in open
grates are burning, is not perfect. If only two or three
persons are living in the room, and a large fire is kept
burning, so much air passes through the room that the
chief part of the noxious vapor is drawn off by the current and the influence of general diffusion; but if there

are many persons in the room, or the fire is small, this will not be the case. All that part of the atmosphere of the room that lies below the throat of the fire-place will then flow steadily in a continuous stream towards the chimney; but the light impure vapors will collect in the more stagnant air above to a very injurious extent. The most efficient way of remedying this evil consists in the introduction of Dr. Arnott's ingenious balanced valve through the upper part of the wall of the room into the chimney shaft; then the impure air floats out partly through the fire below, and partly through the valve above, so that the impure accumulations are drawn off. In order that Dr. Arnott's valve may act efficiently, it is necessary, however, that the throat of the fire-place should be narrowed considerably more than it usually is. If this is not done, there may not be a sufficient supply of fresh air through the crevices of the doors and windows to keep a steady outward current through both the fire and the valve; and, consequently, the fire will draw upon the valve for a part of its supply, and so keep it shut and inoperative, bringing a downward current with a little smoke, through it, as it closes, now and then.*

743. Even when only the purest air is breathed, *brisk exercise must be sometimes*

---

* Messrs. Bailey, of 272 Holborn, now manufacture very convenient *circular* ventilators, each adapted to a cast-iron tube, for building into a hole made in the walls, which they sell for 6s. 6d. All that is requisite to give control of the most troublesome draught, and make these very ingenious valves perfectly operative, is the construction of a low arch of brick-work, from the hobs of the ordinary fire-grate, entirely closing in all the rest of the front opening, between the fire and the chimney. The crown of the arch should not be more than four or five inches above the upper bar of the grate, and the passage into the chimney shaft above should be a square opening about five inches long each way. By this simple and inexpensive proceeding, any draughty and smoky room may be converted into a warm and comfortable apartment. The low arch quickens and steadies the draught like a blower; and the narrow opening above, into the chimney shaft, prevents the consumption of fuel from becoming too great, and the fire too fierce, in consequence of the quickened draught. The addition of an adjustible damper-plate to this opening gives the power of increasing or diminishing the draught.

taken in order to render the purifying work of respiration complete.

When brisk muscular exercise is taken, the blood is caused to move more quickly through the channels of the circulation (502). If it moves more quickly through the capillary vessels of the lungs, it throws out more carbonic acid and other exhalations into the air cells in any given time. Air that is just expelled from the lungs during brisk exercise, contains one-third more carbonic acid, than air that is expelled during repose, and this continues to be the case for about an hour subsequently to the exertion. If the exertion is pushed till great fatigue is experienced, however, the exhalation of carbonic acid is less. Exercise thus blows up the internal furnace of the animal frame, and quickens all those vital operations and purifying actions in which the chemical energies of oxygen are concerned. When people take no exercise, but remain lounging or sitting day after day, the blood is sure to get loaded with impurities, from which it would be relieved under better habits.

744. The main object of the process of respiration is thus of a double nature ; the *supply of oxygen* to the blood, and the *removal from the blood of redundant carbon, and of other more volatile impurities* in a lesser degree.

The air tubes of the lungs thus act at once as the blast-pipe, and the chimney of the organic frame. They supply the stream that fans the vital activities of the economy, and they carry off the products of the vital changes, as chimneys carry off vapors and smoke from burning furnaces.

745. When air is entirely excluded from their lungs, animals speedily *die of suffocation.*

The carbonic acid, and other impurities, that are injurious to the system, are then kept in the blood, and are caused to circulate with it again and again through all the recesses of the body. About three additional grains of carbon are added to the blood every minute that the purifying process is checked. This, however, cannot be continued long. The dark venous blood, that is transmitted to the lungs, is sent on to the left side of the heart (424), not as bright arterial, but as still impure venous blood. This venous blood is then pumped on to the muscles, to the brain, and to all the active organs of the system ; but in them it acts as a poison, in the place of a stimulus and a food. In five minutes, all powers of voluntary motion and feeling are lost. In another five minutes, the heart ceases to beat, and the circulation of the blood stops altogether. This result is most commonly seen to take place in consequence of the immersion of the body in water, when air is prevented from reaching the cavities of the lungs by the intervention of this liquid, or in consequence of the opening of the wind-pipe being violently closed by hanging, or by some accidental injury or other chance occurrence. The word " *suffocation*" which is used to designate the result of this forcible exclusion of air from the chest, is taken from the Latin " *suffoco*," which means " *to strangle*," or " *stop the breath*," by something placed beneath the jaws (*sub and faux*).

746. In cases of suffocation, the heart *retains its contractile power some little time* after it has ceased to act.

So that if the natural stimulus of arterial blood is restored to its left chambers, it may resume its activity, and all the suspended powers of animation be re-awakened in the frame. In all cases of apparent death from drowning or suffocation, warmth should be applied to the surface, and the skin should be freely rubbed in the hope of causing the blood to recommence its movement through the capillaries. Fresh air should be

allowed at the same time to blow freely upon the face,
in order that it may find ready passage into the chest,
in case any movement should be produced in the walls
of the cavity under the influence of the renewed motion
of the blood. It is even possible to continue the play
of the chest artificially for some time, and so to keep
pumping fresh air into its air cells. The movement of
the blood in the lungs is partially carried on by the
mutual relations existing between the walls of the vessels, the liquid itself, and pure air contained within the
air cells; so that if pure air be forced continually into
these chambers, the stagnant blood in the capillaries
may be arterialized sufficiently once more to begin to
travel onwards through its channels. If its journey be
continued until it reaches the left chamber of the heart,
the arterial blood immediately excites the dormant contractility of the walls of that organ, and so the heart
begins to beat again, all the vital functions are resumed,
and the expiring spark of life is fanned into a flame.
There have been many cases in which life has been restored by the assiduous use of artificial respiration,
after all its active signs had been entirely suspended
for as much as half an hour.*

747. The habitual breathing of impure air is a kind of slow suffocation.

When air is breathed that is already overcharged

* Every one ought to be aware how easy a thing it is to keep up artificial respiration in an apparently dead body. If the walls of the
chest and abdomen be strongly compressed by the spread hands, a
considerable quantity of the impure air contained in the air-cells of
the lungs is immediately squeezed out. If then the nozzle of a pair
of bellows be introduced into one of the nostrils, and the skin be
carefully closed around it, the other nostril and mouth being held
firmly shut, and the prominent cartilage of the throat (the larynx)
being at the same time strongly *pressed back* to close the swallow and
prevent the air from finding its way into the stomach, when the bellows are worked, fresh air will be blown into the cells. By alternately blowing with the bellows, and pressing by the hands (one
nostril being then left open), the natural action of respiration may
be imitated by the hour at a time. About three pints of air may be
safely squeezed out and blown in at once in this way, to a person of
full size.

with carbonic acid, the consequence is that less than the
usual amount of this compound gas is thrown out at
each breathing. This is so much the case that when the
air that is taken into the chest contains one or two per
cent. of carbonic acid, it is found not to have so much as
five per cent. when thrown out again (721). When impure air is habitually breathed, therefore, the blood is
kept in an impure state from the accumulation in its
streams of carbonized matters that ought to be removed,
and so the entire body is slowly poisoned by that which
ought to be its food.

748. The blood in the capillary vessels of
the lungs *absorbs any kind of vapor that is
introduced into the air cells in process of
breathing.*

It has been seen how oxygen is sucked into the blood
through the walls of the pulmonary capillaries (710).
When the air that is breathed is impregnated by any kind
of vapor not usually present in it in nature, portions of
that vapor, too, are sucked into the blood in the same
way. It is by this means that the influence of chloroform, in blunting or destroying for the time the sensibility of the body, is secured. When the air in the air
cells of the chest is kept loaded with the volatile vapor
of this agent, portions of it mingle with the blood, and
circulate with it through the system. When it reaches
the brain, it arrests its activity, and in some unknown
way prevents its vesicular substance from performing
its usual offices, that of perceiving sensory impressions
amongst the number. When sensations cannot be felt,
pain, which is merely intense sensation (628), is also
unheeded. Hence, the inhalation of chloroform is used
for the prevention of pain, when severe surgical operations are performed. Just as the vapor of chloroform
is introduced into the blood, any other kind of poisonous
vapor, with which the air is charged, may be communicated to it. It is in this way that the impure exhalation of the breath proves injurious when it is allowed to
accumulate in badly ventilated rooms where people dwell.

CHAPTER XXIV.

THE SKIN.

749. The external surface of the animal body is covered over by a stout layer of membrane, which is called *the skin.*

("*Skin*" is derived from the old Anglo-Saxon word "*scinnan,*" "*to shine.*") The skin is composed of fibres, interlaced and firmly woven together, until a very stout membrane is formed (437). This membrane is of a very dense and close texture externally, but its under surface is comparatively loose, and divided by interstices like those of *connective tissue* (436). The skin and connective tissue, indeed, pass insensibly into each other. It is in this way that the skin is held firmly, although loosely, upon the parts to which it is applied. Fat cells are generally stored abundantly in the loose interstices of the under portion of the skin (451).

750. The external surface of the skin is *raised into a series of ridges* that have furrows between them.

The ridges of the skin are most plainly seen on the ends of the fingers. They are made up of rows of little projecting papillæ. These are the papillæ that have been already described as constituting the organs of touch (626).

751. The ridged surface of the skin is fur-

nished with an outer covering that is altogether *insensible and devoid of vitality.*

Neither nerves nor blood-vessels enter this outer covering of the skin. A fine needle may be pushed through it without either drawing blood, or causing pain. This external insensible layer is intended as a defence to the acutely sensitive structure that is placed beneath. If it were not for its presence, the contact of every material substance would cause pain instead of sensation. When blisters are applied to the skin, it is made by the consequent inflammation to throw off this outer defence, and then the touch of even the thin soft air causes intense smarting. If there were no outer insensible layer to the skin, men would always feel 'all over as they do when blistered by fire, or other like destructive agents.

752. **The outer insensible layer of the skin is called the** *cuticle,* **or** *scarf skin.*

("*Cutis*" is the Latin word for skin. "*Cuticle*" therefore means "*the little skin.*") The thickness of the cuticle varies in different situations, according to the amount of protection that is required, from a tenth to a two-hundredth of an inch. ("*Scarf*" is taken from the Anglo-Saxon "*scearp,*" "*clothing,*" or "*apparel.*")

753. **The cuticle is composed of** *successive layers of dried and flattened cells.*

These cells are originally formed upon the surface of the true and sensitive skin. At the time of their formation they are soft vesicles filled with fluid, but they are pushed forwards towards the outer air in layers, as new vesicles appear beneath, and as they are thus advanced, get more and more dried by exposure to the atmosphere and evaporation, until at length they are merely flattened scales entirely devoid of life (466). These scales of the cuticle correspond with the cell layers of the mucous membrane (443). By this arrangement there are always masses of delicate soft cells in immediate contact with the sensitive papillæ of the true skin, and y*t only dry hard resisting scales are exposed to the air, and the rough contact of external bodies. In *Figure* 90, this structure of the skin is represented.

Fig. 90.

p p are the projecting sensitive papillæ of the true skin. *v v* are the young vesicles of the cuticle lying in contact with and between the papillæ; and *c c* are the external layers of dried scales.

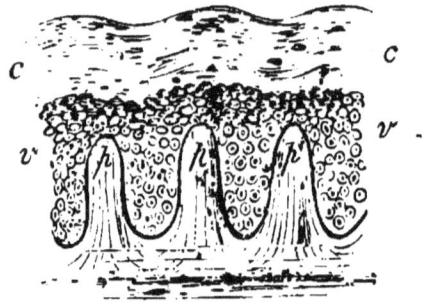

754. The scales of the cuticle *are continually being cast off* from its outer surface.

New layers of vesicles are constantly in process of formation on the external surface of the true skin, and the old layers are as constantly being raised, or pushed outwards by their production until they come in contact with the air, when they in their turn are brushed off or shed. This successive shedding of the cuticle is one of the means designed for keeping the skin continually in a serviceable state for the delicate work it has to perform.

755. The *substance* of the cuticle *is impervious to moisture or vapor.*

The dry scales are so closely packed together, and there are so many layers of them, that vapor cannot easily find its way through among them. It is altogether essential that the external surface of the body should be furnished with some covering of this kind that is capable of preventing evaporation, as otherwise it would be always reeking with steam thrown off from the warm liquids circulating in the delicate vessels within.

756. But *the cuticle is pervious,* although the cuticular substance itself is not so. Its layers are *pierced* by a great number of *holes,* or *pores.* ·

In *Figure* 91, a sketch is given of the appearance which the outer surface of a small piece of cuticle presents, when considerable magnified by the microscope. The openings of the little pores by which its thickness is traversed, are shown, lying between the spaces that correspond with the points of the papillæ of the sensible skin beneath. In some parts of the human skin, there are as many as three thousand of these pores upon each square inch of surface. It has been calculated that the human body contains no less than three millions of them altogether.

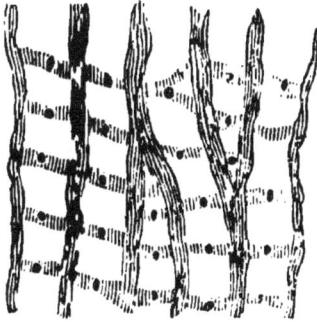

Fig. 91.

757. The pores of the cuticle are *the mouths of little tubes*, that come from within.

A little tube dips in from each pore through the successive layers of the scarf skin, and through the fibrous substance of the true skin beneath, and then it is coiled or twisted up on itself into a sort of knot, which lies snugly in the interstices of the loose connective tissue (749). This arrangement of the twisted tube, and its passage through the substance of the skin to the external cuticular pore is represented in *Figure* 92. Each tube is about half an inch long when its knotted portion is unrolled. If all the 3 millions contained in the skin of a man were stretched out, and laid end to end, they would form a continuous pipe twenty-eight miles long ! *The human skin*

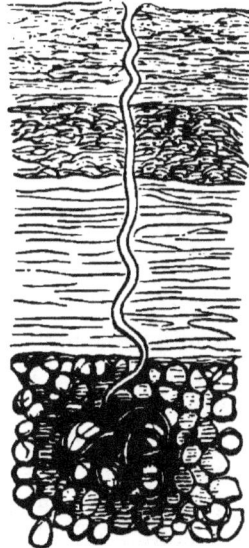

Fig. 92.

has twenty-eight miles of delicate piping connected with
its structure, for the performance of a service of great
importance to the frame.

758. Each tube of the skin has *an abundant supply of blood-vessels* furnished to its
knotted part.

Capillary blood-vessels are supplied to the walls of
the twisted tube, much in the same way as they are to
the walls of the air cells of the lungs (707). Streams
of blood are thus kept constantly flowing round the cavities of the tubes, with nothing but the delicate membranes of the vessels separating the liquid from their
open spaces.

759. The tubes of the skin are *filled with
vapor*, which is exhaled into them from the
blood that courses through the capillary
vessels, distributed to their walls.

This vapor streams out from the warm blood, through
the delicate films of the capillary vessels, into the cavities of the tubes. The evaporation that would take
place over the entire surface of the skin, if there were
no layers of dried cuticular scales to prevent it, really
does take place into little hollow caverns, made ready
for its reception beneath the skin (764).

760. The vapor, that is exhaled from the
blood into the tubes of the skin, *finds its way
into the outer air through the pores* of the
cuticle (756).

It would not have been sufficient for the purpose of
evaporation, if the little openings had been left through
the cuticle, without any extended cavities having been
connected with them beneath. This will at once be understood when the statement, made above, is remem:
bered, that there is as much as 28 miles of this little   •
tubing contained altogether in the skin of a full-grown
man. The consequence of this plan is that an evapora-

ting surface is furnished considerably larger than the
entire skin would have been, if it had been all pores.

761. **The vapor, that passes out from the
body through the pores of the skin,** *is called
the perspiration.*

("*Perspiration*" is derived from two Latin words
"*per*" and "*spiro*," which signify "*through*" and "*to
exhale.*") The perspiration is that which is exhaled
through the skin. In a general way the perspired vapor
escapes from the pores of the skin as invisible steam; it
is then called *insensible perspiration.* When, however,
it is poured out from them very quickly, it is condensed
into drops of liquid as soon as it comes into connection
with the air, and moistens the external surface of the
skin.

762. **The quantity of perspiration, that
escapes through the pores of the skin,** *varies
exceedingly* at different times.

Never less than a pint of water is exhaled from the
blood through the pores of the skin in twenty-four hours.
Sometimes as much as four pints are thrown off in the
same time.

763. **The body is** *cooled by the evaporation
that takes place through the pores of its
skin.*

If a piece of wet linen is laid upon the skin, and left
there for a short time exposed to the air, its moisture
evaporates from it, and produces a sense of coldness in
the skin as it does so. The heat of the skin is used up
in converting the water, that had been soaked up by the
linen, into vapor or steam. The steam flies away with
the warmth of the skin, and consequently the skin feels
chilly and cold. Exactly in the same way that the evaporation from a piece of moist linen cools the skin, the
evaporation from the blood through the pores of the skin
cools the body. Perspiration is thus directly opposite
in its influence to respiration. Respiration blows up the

slow furnace contained within the frame, and so heats it
above the temperature of the surrounding air. Perspiration carries away portions of the heat thus generated,
and so helps to cool it down again. The vapor that is
exhaled through the lungs (723) plays its part in this
cooling process; the pulmonary exhalation is indeed but
a modification of that of the skin. Some animals, like
the dog, lose nearly all their exhaled vapor through the
air cells of the lungs, instead of through the skin; hence,
these animals open their mouths, and pant whenever the
exhalation is very greatly increased. In some circumstances, more than five pints of liquid are poured out
through the human skin and lungs together in twenty-
four hours.

764. The perspiration is increased, whenever the *body is made too hot.*

The internal furnace of the body consumes pretty
nearly the same amount of fuel at all times, and consequently forms pretty much the same quantity of heat at
one time as at another; but that body is placed in very
varying temperatures at different seasons. The heat of
the air during the winter in England is sometimes ten
or more degrees below freezing water for days. During
the summer it is sometimes fifty degrees hotter than
freezing water. Men, too, live for long periods together
in the cold frozen regions of the earth, where the temperature of the air is occasionally eighty degrees colder
than frost. In India a temperature one hundred degrees
hotter than frost is sometimes endured. All this variety
it was intended that man should be able to bear. Consequently while the internal production of heat has been
made pretty steady, means have been furnished him
whereby he may economize this heat on occasions when
it is hardly enough for his wants, and whereby he may
let it off to waste under circumstances when less of it is
thrown of by the more ordinary process of cooling from
simple exposure. The pores of the skin are the cooling
apparatus of the body, whereby its superfluous warmth is
allowed to escape with rapidity, whenever at any time

it threatens to accumulate sufficiently to overburden the frame. It is on account of this need for a regular exhalation according to the circmstances of the time, that the evaporation from the blood is thrown into little cavernous spaces beneath the skin, instead of being poured off at once into the outer air (759). This is the meaning of the ingenious apparatus of perspiration tubes, and cuticular pores. The caverns, into which the vapor is first exhaled, have little outer doors, which can be opened or closed, according to the amount that is intended to be passed through them. When the temperature of the air is low, and the body needs all its warmth, the pores of the cuticle are closed by the chilling effect of the external cold, which shrivels in and presses down the surrounding scales (756). But when the air is very warm, and the body becomes oppressively hot, these pores are relaxed and opened, the blood is made to rush more rapidly through the capillaries spread upon the walls of the perspiration tubes, and streams of vapor are poured rapidly out into the atmosphere, carrying more and more heat with them as long as they continue to flow.

765. The most pleasant temperature in which the body can be placed, (so far as the warmth of the external air is concerned,) is *sixty-two degrees* of the heat scale.

Impressions of warmth and cold are merely sensations caused by the more or less rapid removal from the sensitive surface of the body, of the heat generated in its own interior furnace. When the heat is carried off by the surrounding air more rapidly than it is supplied from within, the skin is sensible of a chill. When heat comes to it from within more rapidly than it can be dissipated into the surrounding air, the skin feels hot. The internal portions of the body are kept at a temperature of about 100 degrees of the heat scale during health, and the skin is a little colder than this. But the substance of the air is so slow a conveyor of heat, that its contact produces no feeling of chill even when it is many

degrees colder than this, unless it is kept moving over
the body in a rapid stream. When the air stands at a
temperature of 62 degrees, without much motion, heat
is carried off by it from the skin just about as rapidly
as it is communicated to it from within. Consequently,
in such a temperature, no external aid is needed to preserve either the warmth or the coolness of the skin.
When air stands at 55 degrees, light clothing and gentle
exercise just serve to keep the skin at the degree of
warmth that is most agreeable to its own sensations.
Hence, the atmosphere is said to be *temperate* when it
possesses this amount of heat. (The word " *temperate*"
is derived from the Latin " *tempero*," " *to regulate*," or
" *make soft*.") The body feels colder when exposed to
a wind, the air composing which is at 62 degrees, than
it does in still air at 55 degrees, because the current
constantly brings fresh aerial particles into contact with
the skin, each having the same capacity for the reception
of heat from it. So also it feels comfortably cool in air
that is at 70 degrees, if that air is moving past it in the
form of a brisk breeze.

766. Clothes keep the body warm *by re-
tarding the escape of its heat* into surrounding
air.

Clothes retard the escape of heat in two ways. In the
first place, they prevent currents of air from blowing
against the skin, and so robbing it of heat received from
it during the instant of contact. In the second place,
they prevent heat from being thrown off so quickly by
its own innate tendency to escape, or from being radiated
(rayed off) as this is called. In order that clothes may
be able to effect these purposes (and more especially the
latter one) it is necessary that they should be made of
substances that have very little power of carrying heat;
or, in other words, that do not allow heat to run along,
or travel, through them. Skin, that feels cold when exposed to the air, becomes comfortably warm when covered by flannel or woollen cloth, because the heat, that

is sent to it from within, is then prevented from escaping any further, and is so held in the skin for the time, and employed in the service of communicating to it an agreeable sensation. Clothes keep the body warm, because they husband the warmth that is supplied to its outer surface from the internal furnace, instead of its being allowed to fly off in waste as rapidly as it is received.

767. Hair is the *natural clothing* of the body.

Most species of animals have warm outer coats provided for the protection of their skins from chilling exposure to the air. Whenever this provision is needed, the scales of the external cuticle (753) are produced more abundantly, and are piled into little flexible cylinders, which are thickly planted, side by side, instead of being merely arranged in shallow layers, somewhat after the fashion of the tiles of a roof. These cuticular cylinders, added to the skins of animals when they are to be made into great coats, are called " *hairs*," (the word being a slight modification of the old Anglo-Saxon "*hær*.")

768. The head is the only part of the human body that is furnished by nature with an *external hairy coat*.

This additional care is taken of the head on account of the very important and delicate organ that it contains. The brain is of so frail a texture, and of such exquisite organization, that it cannot bear great sudden changes of temperature with impunity; hence, the skin, that covers the bony case in which the organ is lodged, has an outer defence of long thick hair placed nearly all over it, in such a way that the natural warmth must be slow in escaping from within, even when the head is exposed to great external cold, and that great heat must be equally hindered from finding its way in. Closely packed hair is of such a nature that it does not allow heat to travel through its substance otherwise than very slowly.

769. Hairs *are produced by little vascular papillæ placed in small pits* hollowed out in the substance of the skin.

Hairs are enlarged at their bottoms into what are termed their bulbs. The bulb of a hair is lodged in a pit hollowed out in the fibrous substance of the skin: this is called the "*hair follicle*," (from "*folliculus*," the Latin term for a "*little bag*.") The means, by which the ordinary scale layers of the cuticle are changed into cylindrical hairs, are very curious and beautiful. *Figure* 93 will help to make the plan easily intelligible. *f f* represents the little bag or hair follicle. This is formed of the ordinary fibrous substance of the true skin. At the bottom of the follicular bag a little pimple or "*papilla*" (*P*) projects into the cavity. This papilla is abundantly supplied with blood-vessels, and is indeed the organ mainly concerned in the formation of the cells out of which the hair is made. The cuticle is prolonged down into the bag, and its young cells entirely surround both the papilla and the bulb of the hair. *c c* are masses of cuticle cells. When a hair is formed, portions of these cells are pushed up, as new ones are made beneath, along the middle of the follicle, and are enclosed in a sheath of dry hard scales, set obliquely upon each other, and over-lapping, as shown at *s s*. These form the outer covering of the hair. Within

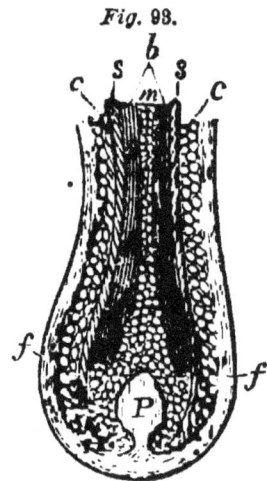

Fig. 93.

this sheath, the cells get compressed, as they are pushed upwards, into closely packed fibres, which constitute what is called the *cortex*, or bark of the hair (*b*). Within the bark there is a portion that is still filled with cells, somewhat compressed, but not changed into fibres (*m*); this is called the "*medulla*," or pith of the hair. The layer of cuticle scales (*c c*) that is in immediate contact

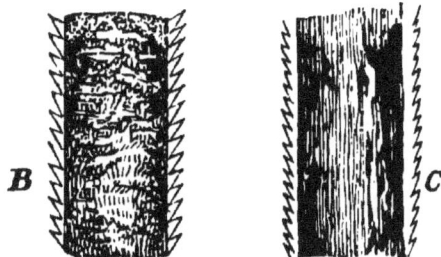

with the sheath of the hair (*s s*) is dry like the ordinary outer surface of the skin, so that the hair cylinder is easily pushed upwards within it, as it grows. A hair thus consists of a hard sheath, a firm bark, and a loose pith, all formed from ordinary cuticle cells, framed by a little vascular pimple sunk in the skin, and pushed out through a hollow tube of the cuticular substance, so arranged as to allow of the passage. These various parts are easily observed in the porcupine's quill, which is nothing but a gigantic hair. In the hair of the human head, the cells are nearly all condensed into fibres, within the sheath, so that there is no pith left. *Figure* 94 shows the appearance of the human hair when immensely magnified. The projections at the edge and the transverse markings are due to the over-lapping scales of the external sheath (*s s, Fig.* 93). The sketch at *C* shows the appearance of the human hair when it is split through the middle; it is then to be entirely composed within the sheath of closely-packed fibres without any loose pith.

Fig. 94.

770. **The human skin has been left naked by nature, because man has intelligence, and** *is able to provide himself clothing* **by the** **exercise of his ingenuity.**

When man needs warm clothing, he takes down from the cotton-grass, and threads from the flax-plant, and borrows wool from the sheep, and he weaves these materials into convenient fabrics that he can make thick or thin, and wear or throw off at pleasure. Man is able to bear the heat of equinoctial climates much better in consequence of not having his skin burthened of necessity

22   O

with any hairy covering; and when he goes into the temperate or frigid regions, he takes care to provide himself with defensive garments; so that the nakedness of his skin is a positive advantage in the matter of adaptability to circumstances.

771. Man is able to exist with safety and comfort in *extreme degrees of atmospheric temperature*, both of heat and cold.

In the frigid regions near the Pole, the air is sometimes 102 degrees colder than freezing water; in India it sometimes becomes 130 degrees hotter than freezing water, in the shade of canvas tents : yet man, nevertheless, manages to live in both these regions. No other kind of animal, except perhaps the dog, exists over so wide a range of the earth's surface. Even in what are called temperate climates, the air is sometimes 110 or 120 degrees warmer than it is at others. Man is able to meet these great variations of external temperature, without suffering injury from them, in consequence of three things : he is able in the first place, so to modify his diet as to throw more fuel into his blood when he needs it (311) ; in the second place, he can husband the heat that is formed within his body by putting on thick clothing (766), and by keeping the interior of his dwelling warm by means of artificial fires; or, in the third place, he can cool himself when excessively hot by throwing off some of his clothes, and by drinking abundance of water, in order that it may be turned into perspiration.

772. The *direct application of cold* to the skin *is depressing* in its influence.

When the skin is exposed to great cold, it shrinks ; the circulation of the blood through its capillaries is either diminished, or checked, and the nerves of its papillæ are rendered less sensible.

773. The *subsequent influence of cold* applied to the skin (if not very severe) is *stimulating and invigorating.*

When the circulation of the blood is checked in the capillaries of the skin, the heart becomes sensible of the increased resistance, and tries to overcome it by throwing increased force into its own work. If the cold be not very severe or even if it be so, but only continued for a short time, the heart succeeds in its attempt; the state of languor and chill passes into one of comparative excitement; the blood flows more freely than usual through the skin, and the numbness and chilliness there are exchanged for a sense of agreeable glow. Cold, when encountered in moderate intensity, or when, if severe, counteracted by the influence of brisk exercise and warm clothing, is rousing and strengthening to the powers of life.

774. The *direct application of heat* to the skin is *stimulating in its influence.*

When the skin is exposed to great warmth, the blood is at once caused to move with increased speed through its vessels, the nerves of its papillæ are roused to augmented sensibility, and perspiration is abundantly poured out through its pores.

775. The *subsequent influence of great heat* applied to the skin is *depressing and debilitating.*

No vital activity can be maintained in a state of undue excitement for prolonged periods of time. The activity of the present must be, sooner or later, followed by languor and exhaustion. It is, so to speak, taken out of the power which should belong to the future. Any organ that is made to exert itself beyond its ordinary capacity for work, must in a short period have its energies exhausted by the effort. Hence, it is that, in warm climates, or very hot seasons, weariness, languor, and unwillingness to make exertion are soon experienced. Most of the vital organs, in common with the skin, are at first stimulated to increased activity by the heat, and subsequently depressed and exhausted by their efforts.

776. The continuance of severe cold soon *destroys all vital powers.*

If sufficient artificial means are not brought into play to maintain the warmth of the body, when severe cold is long encountered, the circulation flags more and more; first the skin, and then other parts lose their powers of feeling, all the actions of life become more and more languid, and at last death ensues.

777. *Internal organs are overloaded with blood,* when cold is applied to the skin.

The blood that is prevented from entering the capillary vessels of the skin, on the application of cold to them, must go somewhere. At first it hangs back upon the large vessels; but the increased force that the heart puts on, upon becoming sensible of the resistance, soon drives it out of them. It then flows in greater quantities to all those internal organs, to which the chill does not reach, and in particular to the various tracts of mucous membrane that line the internal cavities.

778. *Colds are caught in consequence of the overloading of internal organs,* by a chill applied to the skin.

If after the skin has been chilled, it is soon brought into a state of glow by reaction, the internal organs that have been overburthened with blood, are at once relieved, and no mischief ensues; but if, on the other hand, the skin continues to be cold and inactive, the overloaded organs become excited by the increased work that is demanded of them in order to dispose of the augmented supply of blood. Most generally they break down in the attempt to do more than their usual amount of work, and have their own natural actions deranged. This derangement leads to the various feelings of discomfort that are together known under the familiar denomination of "a cold in the head," or "a cold in the chest," or some other similar kind of disorder.

779. Disorder is produced in the system,

not by the influence of extremes of temperature, but rather by *sudden changes from one
kind of extreme to the other*.

Colds, and other like kinds of disorder are not produced by simple exposure to cold; such simple exposure
only calls up reaction in the healthy frame. Heat, or
great fatigue is first encountered, and then a sudden chill
is experienced. Heat first depresses and exhausts the
vigor of the vital organs, and saps their powers of
resistance and reaction, and then some sudden chill
to the skin occurs to complete the work. When a
delicate structure, that is in a relaxed and weakened condition in consequence of having been for some time
stimulated to over-work, suddenly has an undue load of
blood sent to its capillaries, the greater portion of which
ought to have gone to those of the skin, the weakened
parts are altogether unable to set up enough reaction to
free themselves from the oppression, and so stagnation
in their own channels, and disorder very speedily follow.
Mischief of this kind occurs more constantly when a
change of weather comes after long continued heat, than
in very cold weather; because the vital organs are then
all in a debilitated state in consequence of the excitement
they have been suffering during the prevalence of the
high temperature, and are so predisposed to suffer if
they get overburthened with blood in consequence of
the occurrence of a chill.

780. Exposure to chilling influences should
be chiefly guarded against *when the body is
in a state of exhaustion* or fatigue.

There are two most important facts, that every one
ought to bear constantly and carefully in mind. Danger
to health lies in exposure to sudden changes of temperature, and not in exposure to heat or cold, taken by themselves: and that exposure is perilous in proportion to
the exhaustion or fatigue of the body, at the time when
it is encountered. Any one who has been refreshed by
a long night's sleep may jump up from a warm bed, and

plunge into a bath of the coldest water without risk to his health, who would nevertheless be sure to suffer some serious disorder if he did the same thing when heated and fatigued by long-continued labor or walking. There is no danger in jumping into cold water when the body is profusely covered by perspiration, provided it is at the time vigorous and fresh; but if the same thing is done, when the body is weary, as well as perspiring, some grave internal mischief will be almost sure to follow. Sudden exposure to cold is the most common of all the influences that call up disorder, in frames that have been already predisposed to become deranged by the operation of other causes. The precise nature of the disorder that follows depends upon the circumstances, in which the person who suffers it, has been living. That particular organ is attacked, or suffers most, which chances to have been previously excited and weakened the most. Ordinary colds are nearly always caught on leaving warm rooms at night after encountering excitement or exertion for some continued length of time.

781. The perspiration contains a *small quantity of solid matter*.

By far the larger part of the perspiration is simply water; but there is a small quantity of a less pure ingredient mingled with it. This is something of the same nature as the impure exhalation that is conjoined with the vapor of the breath (724). It is an ammoniacal compound (119) formed from the decay of the albuminous principles of the organized structures. As much as 100 grains of this decaying substance are commonly thrown out of the body in the perspiration every 24 hours, and sometimes very much more. The skin is thus an auxiliary to the lungs in the work of purifying the blood. A small quantity of decayed noxious matter, that has completed its work in the system, is continually being got rid of through the perspiration tubes and pores.

782. The skin is *kept soft* by means of an oily liquid that is formed for the purpose.

In consequence of the constant exposure to which the skin is subjected, as constituting the external surface of the body, it would soon get so dry and cracked as to be unfit for any of the purposes for which it is designed, if some means were not adopted to prevent the result. In order that this may not happen, it is in reality anointed carefully by a sort of oil that is pressed out from the hair follicles (769). Clusters of little secreting cells of a peculiar kind are placed near to the sides of the pits that contain the bulbs of the hairs, and throw the secretion which they separate from the blood into the pits, through small tubes that run to them to serve as channels of conveyance. These clusters of secreting cells are called "*sebaceous glands*" (447). They are so termed on account of the oily liquid that they form. ("*Sebum*" is the Latin word for "*tallow*" or "*suet.*") *Figure* 95 represents the appearance of a pair of these glands sending their tubes into the cavity of a hair follicle, *B* being the bulb of the hair. Those parts of the skin that are most exposed to the drying and cracking influence of the sun and air, as the head and face; and those parts that are so placed as to be most subjected to rubbing, as the folds near the joints, are the most abundantly furnished with sebaceous glands, and are the most freely anointed with their oily production.

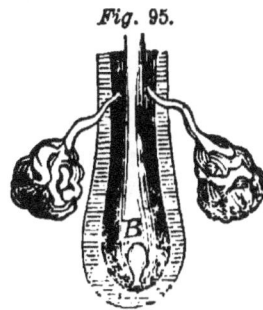

Fig. 95.

783. The skin is thus a very complicated structure, *adapted for the performance of a variety of offices.*

Besides being a dense fibrous membrane suited for the preservation and defence of the parts placed beneath, it has nervous papillæ to enable it to feel (750); a thatch ing of overlapping scales, to protect the papillæ (753); oil-glands to keep these scales duly dressed, so that they

do not become so dry and horny as to be unmanageable
(782); hairs, that can be made into a warm great coat,
when they are needed (769); caverns for the reception
of vapors exhaled from the blood (757); pores, which
are the adjustable openings of those caverns (756); and
besides all these, an abundant supply of capillary blood-
vessels, which carry blood to all these several parts,
with the exception of the scales of the cuticles, and the
perfected cylinders of the hairs.

784. The pores of the skin are *kept open by
the shedding of the scales* of the cuticle.

The decaying organic matter contained in the perspiration (781), and the oily secretion thrown out from the hair-
follicles (782) mingle with the dust and dirt that necessarily come in contact with the outer surface of the body,
and so form a thick tenacious deposit, which would soon
choke up the pores, and prevent exhalation from them,
if some means were not taken for its removal. Nature's
plan of removing the soil of the skin is the successively
throwing it off, with the immediate outer surface (the
external cuticular scales) upon which it lies.

785. *Bathing assists nature* in keeping the
pores of the skin free.

All the arrangements of the human body have been
made in subservience to the intelligence of the creature.
Operations that are perfected in the economies of the
irrationable animals, irrespective of any voluntary interference on their part, man is expected to complete for
himself. Nature takes care that under all circumstances
a certain amount of cleanliness shall be ensured by the
continued shedding of the surface scales of the skin, but
she intends that her own plans shall be aided by voluntary co-operation, in the frequent use of bathing. She
enables man to understand the organization of his skin,
the offices which it performs, and the properties of water,
and she then allows him to experience the comforts of
cleanliness, when he applies his knowledge. In order
that the surface of the body may be kept in the best

possible state for the healthy performance of its important and varied functions, it is absolutely essential that
it shall be thoroughly washed at least once every day.
When the skin is not kept perfectly free from the accumulation of its own tenacious secretions, as well as from
the accidental deposit of soil from without, its pores are
more or less clogged up, the circulation in its vessels is
obstructed and languid, and its texture becomes harsh
instead of being soft, and its sensation dulled, in the
place of its being highly sensible. But when the pores
of the skin are clogged, and the movement of the blood
in its vessels languid, certain impurities, that ought to be
thrown out from the system by its means, are retained,
to the injury of the internal organs. An imperfect and
deficient action in the skin tends to make the blood poison
ous to other parts of the frame, much in the same way
(although in a less degree) as slow suffocation from the
habitual breathing of impure air (747). Pure air and
personal cleanliness are near allies, so far as their influence upon health is concerned.

786. The *early morning is the best time* for
the employment of the bath.

A bath of the coldest water should be used on rising
from bed in the morning, and this may be advantage-
ously continued all the year round. At this period of
the day the body is fresh, and therefore the least likely
to suffer any mischief from encountering a sudden chill
(778). The transient application of cold to the skin
awakens immediate reaction, and so produces an invigorating effect upon the system at large (773). But it
produces a further benefit than this, it hardens, as well
as invigorates; that is to say, it accustoms the skin, so to
speak, to the influence of cold, and educates it to rise at
once into an antagonistic glow. Persons who use cold
bathing on rising from bed all the year round, are far less
liable to catch cold, and to suffer from the various disorders to which this is apt to lead, than those who do
not. It is a mistake to employ cold bathing as a refreshment when the body is in a state of fatigue, for then the

o*

danger is greatest that reaction may not follow upon the effects of the chill. Rest should come first, and bathing afterwards.

———————◆———————

CHAPTER XXV.

———————

DRINK.

———————

787. Nearly all the vital operations of the animal body *are performed* upon, or are aided by the presence of, *liquid substance.*

The food is digested or liquefied before it is taken into the vessels of the body (281). The blood is liquid in order that it may be easily poured through the channels of the circulation (378). The secreted juices destined for specific service within the system, are liquids (447). The contents of active vesicles are all liquid in order that the movements and changes of osmose may be established (170). Liquidity is an essential condition to the greater part of the material that is doing the work of life.

788. *Water* is Nature's great *agent in the production of fluidity.*

Water is the only liquid that is naturally present in any quantity upon the surface of the earth. It is this limpid fluid that is employed in rendering the materials used in organization mobile, and fit for the purposes of life. Water is peculiarly adapted for this work, on account of its neutral character (77). It is able to dissolve away a great variety of compounds into its substance, and there to mingle them so intimately that all their mutual influences and re-actions may be brought into play (379), and yet it is itself so indifferent and bland that it does not materially alter or interfere with the properties of the matters it holds in solution. It is very ready to undertake the charge and transport of whatever may be entrusted to it; but it behaves alike impartially and honestly to every charge, bearing its molecules carefully along amidst its streams, but rendering them up, whenever required to do so, unaltered in all their important qualities. . .

789. *Two-thirds of the animal body* consist of pure water.

The great value of water, in the economy of life, is shown in the fact that by far the larger portion of the structures of animals is entirely composed of it. It has been seen that it forms nearly one-fifth of the blood (377); the brain has rather more water than this in its substance (84 per cent.); the muscles contain three-fourths: even hard rigid bond has one-tenth of water in its composition. In the frame of a man weighing 180 pounds, 120 of the 180 are water, and only the other sixty are divided amongst the remaining matters.

790. Water is continually being *removed from the living body*, during the maintenance of its vital operations.

It has been seen that from two to five pints of water are exhaled through the pores of the skin (762), and through the air cells of the lungs (723), every twenty-four hours. There are also other means by which this fluid is constantly drained off from the system. Water is employed in the removal of waste materials out of the body, as well as in the introduction of nutritious materials into it. It conducts the sewerage operations, as well as those of supply.

791. As water is constantly being drained off from the body, *this waste must as constantly be made good.*

If five pints are taken away every day from the 120

pints which the body contains in its substance, it is clear that these must soon be all removed unless some means are adopted for restoring the loss from time to time.

792. The loss of water, which the body suffers, in the performance of its various vital operations, is *made good by drinking*.

All the water that is thrown off from the body is primarily derived from its circulating liquid—the blood. As this is progressively removed, the blood grows thicker and thicker, until more of the thin fluid is supplied to it. The supply of fresh water to the blood is made through the stomach. Liquids are swallowed from time to time. This act of swallowing liquids is called "*drinking*." ("*Drink*" is a slight modification of the old Anglo-Saxon word "*drinkan*.") The object of drinking is to supply its watery ingredient to the blood, as the object of eating is to furnish it with the various plastic and combustible ingredients (275).

793. Water is taken from the stomach *by the capillary blood-vessels* of the organ.

Water is not absorbed through the same channels as the liquefied food. If this were the case, the supply of dissolved nourishment would be deranged and interfered with whenever the more rapid action of pouring in water was going on. The stomach is lined by mucous membrane which is covered with capillary blood-vessels, coming from arteries and leading to veins. These blood-vessels are only separated from the interior cavity of the stomach by means of very delicate coats, so that whenever any quantity of watery liquid is present in the stomach its water is at once imbibed by the process of osmose, into the blood, and carried on through the veins towards the heart in the ordinary course of the circulation. The absorption of water from the stomach takes place very rapidly. It is not a slow and gradual proceeding like the absorption of digested nourishment by the villi of the bowel (296).

794. Thirst is *Nature's indication* that a fresh supply of water *is needed in the blood.*

Thirst is an uneasy sensation that seems to arise from dryness of the throat, but is really caused by the blood having become too thick for the free performance of its offices. Nature takes this plan of directing attention to the fact, when so much of the water of the blood has been expended that a fresh supply is needed. The thickened blood circulates through all parts of the frame, and through the membranes of the throat amongst them; but in these membranes its presence excites those peculiar sensations of discomfort that impel the creature to drink (321).

795. In *warm seasons and climates more thirst is experienced* than in cold ones.

And consequently more drink is then taken. This is the way in which the cooling of the body is effected, when its internal production of heat is greater than the circumstances of the time require. The pores of the skin are relaxed; the water of the blood escapes rapidly through them as vapor, carrying portions of the superfluous warmth with it: the blood in consequence gets thickened, and thirst is experienced. This causes drink to be taken freely. The increased supply of water is then in its turn rapidly evaporated with still further diminution of heat, and so the oppressed frame is cooled to a sufficient extent.

796. In *cold seasons and climates very little water is evaporated* through the skin.

And so the loss of heat is prevented, which would otherwise be suffered by the body. In cold weather the water that is required to be removed from the frame, in order that certain waste and refuse matters may be transported from it, is poured out almost entirely in the liquid state. This is effected by a very ingenious arrangement of delicate vessels and cells, in the organs called *the kidneys*, which are able to prevent water from

transuding through them out of the blood, when that
blood has a certain degree of thickness, and does not
amount to more than a certain quantity; but whenever
it gets increased in quantity, and thinned beyond a certain extent, the walls of these vessels and cells yield to
the augmented pressure, and water, holding certain saline
matters in solution, passes freely through them. When
the pores of the skin are open, a large quantity of water
escapes from the blood through them; but when the
pores of the skin are closed by cold, the blood-vessels
get more distended by the drink which is taken, and that
increased distension then opens a series of delicate valves,
so to speak, elsewhere, through which it is soon reduced.
But when the water flows off in the liquid state, it takes
comparatively little heat with it, compared with that
which it would consume if it were turned into vapor.
By this contrivance then, water is made to perform the
very important service of regulating the temperature of
the body, besides carrying on the other offices of transporting material, and favoring chemical operations.

797. All the various forms of drink, that
man employs, *owe their dissolving powers to
the waters* which they contain.   •

The many different kinds of drinks, that are in common use among mankind, quench the thirst and restore
the requisite proportion of water to the blood, in spite
of, and not in consequence of, their peculiar properties.
Their efficacy as agents of solution and transport depends
entirely on the quantity of pure water that is present in
them. Every substance, which is already mingled with
that water, serves only to diminish its power of dissolving other matters, and, to this extent, renders it less
valuable in the main business for which it is introduced
into the system. Water, which is nature's chief solvent,
is also the natural and best drink of animals.

798. Fermented liquors, commonly employed as drink, *all contain a considerable
quantity of alcohol* mingled with their water.

Custom and a perversion of the natural appetite, thirst, into a disordered craving for excitement, have rendered the employment of various fermented beverages, in the place of water, very general. These beverages are principally spirits, wine, and beer in an immense variety of forms. Every kind of wine and beer is chiefly composed of water; but also contains small quantities of nutritious and fragrant principles dissolved in the water, and a considerable amount of a peculiar principle called *alcohol*, mingled with it. This principle was known first to the Arabian chemists of the middle ages, and from them received the name it still bears. Whiskey, brandy, rum, and gin, are more than half alcohol, and less than half water. Port and sherry wines contain about one part of alcohol to every four parts of water. Porter and ale contain from one to two parts of it to every 200 parts of water. Small beer consists of one part of alcohol to every thousand parts of water.

799. Alcohol is produced by the *fermentation of sugar.*

All the various fermented beverages are sweet before they are spirituous. Some of the stronger spirits, as rum for instance, are made directly from solutions of sugar. Wine is formed from the juice of the grape, which has a great deal of sugar in it naturally. Beer is brewed from a decoction of malt, which is merely barley that has had its starch turned into sugar by the influence of warmth and moisture. The process, by which sugar is changed into alcohol, is called "*fermentation.*" (The word is derived from the Latin, "*fermento,*" which signifies to "*puff up,*" or "*make spongy.*") Whenever sugar in solution is changed into alcohol, the conversion is attended by an effervescence, or frothing up, in consequence of an occurrence that will be immediately explained (802).

800. The presence of a *warm temperature and of a ferment* is essential to the production of alcohol out of a sweet liquor.

When beer is brewed, the decoction of malt is placed in a warm situation, and yeast is stirred into it. The yeast then sets the sugar of the decoction changing into alcohol, by altering its internal composition. The juice of the grape contains a peculiar substance of an analogous nature to yeast, and therefore alcohol is formed in it, without the addition of any extraneous ferment.

801. When sugar is changed into alcohol, its molecules *lose two-thirds of their oxygen, and one-third of their carbon.*

Alcohol is thus merely sugar with very little oxygen in it, and with a deficiency of carbon. It will be remembered that each molecule of sugar comprises within itself 12 atoms of carbon, 12 of hydrogen, and 12 of oxygen (228). Now, when molecules thus constituted are dissolved in water, and are exposed to a warm temperature, and the influence of a ferment (like yeast), their elements separate from each other and form a new arrangement amongst themselves. One-third of the carbon and two-thirds of the oxygen desert their companions; but the other two-thirds of the carbon, and one-third of the oxygen, remain with the entire quantity of the hydrogen and make with it new compound molecules, which, however, cease to possess the quality of sweetness, and acquire instead the peculiar properties of spirit: that is, when mingled with water, they convert it into a light and inflammable liquid, possessing an agreable penetrating odor, capable of burning like oil when set light to, and of intoxicating animals when taken into their stomachs.

802. The oxygen and carbon, removed from sugar when it is converted into alcohol, *make their escape as carbonic acid gas.*

This, then, is the cause of the effervescence and frothing that accompany the fermentation of sugary liquids. Carbonic acid is set free within them, and bubbles up through them as it escapes into the air. The effervescence of bottled ale and sparkling wines is due to the

influence of the same agent. These liquids are enclosed
in the bottles while still sweet, and when the effervescence
has only just commenced. Consequently as carbonic acid
is set free within them, it is absorbed into the water of
the liquid, and retained there by the force of pressure.
So soon, however, as the cork of the bottle is drawn,
this force ceases to act, and the carbonic acid bubbles up
to the surface. The molecular change that takes place
when sugar is converted into alcohol, may be expressed
to the eye by means of a diagram, thus:

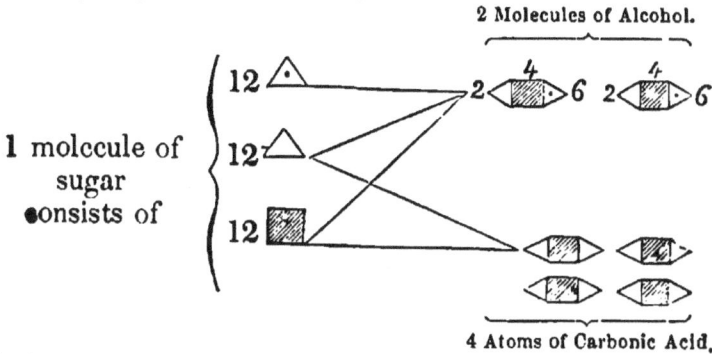

2 Molecules of Alcohol.

1 molecule of
sugar
consists of

12

12

12

2 6 2 6

4 Atoms of Carbonic Acid.

Each original molecule of sugar is split up into two
molecules of alcohol (which remain mingled with the
water), and four compound atoms of carbonic acid
(which bubble off). Each molecule of alcohol therefore
consists of four atoms of carbon, six of hydrogen, and
two of oxygen.

803. The change, by which sugar is converted into alcohol, *is a species of decay.*

It will be remembered that sugar is *built up* by the
constructive powers of vegetable life (228). It is a
highly complex body, made by the combination of several elementary atoms into a compound molecule; but
when this molecule is split or resolved into six simpler
bodies (802), the operation is exactly reversed. That
which vegetable life had *built up,* simple chemical force
pulls down. The fermentation of sugar is the first step
in its destruction or decay. Alcohol is sugar taken one
stage backwards from the complex and organizable state

23

into which it had been raised by the influence of vegetable activity.

804. Alcohol is a much *more readily inflammable* substance than sugar.

Sugar will burn when thrown upon the fire, or when held in the flame of a candle, its carbon being changed into carbonic acid by uniting with the oxygen of the air, and its hydrogen into water by the same means. Alcohol not only does this, but also immediately bursts into a flame when a burning body is brought close to it. Alcohol is one of the easiest substances to inflame of all known bodies.

805. Alcohol is more inflammable than sugar, *because it contains a larger proportion of the combustible elements*, carbon and hydrogen.

When sugar is changed into alcohol, none of the hydrogen is removed. Some of the carbon disappears, but *twice as much of the oxygen* does so likewise. Hence, as there is less oxygen to any given quantity of hydrogen and carbon in alcohol than in sugar, it follows that there must be more hydrogen and carbon to any given quantity of oxygen. In other words, alcohol contains a larger *proportional amount* of hydrogen and carbon, than sugar.

	Carbon.	Hydrogen.	Oxygen.
100 grains of sugar contain, in round numbers . . .	42 grs.	6½ grs.	51½ grs.
But 100 grains of alcohol contain .	52 grs.	13 grs.	34 grs.

(*See note to par.* 239.) 65 per cent. of the substance of alcohol is combustible matter, but only 48 per cent. of the substance of sugar. In sugar, the carbon is already united with half the quantity of oxygen it can remain in union with. Each atom of carbon has already one atom of oxygen associated with it; but in alcohol, there are only four atoms of oxygen to every eight atoms of carbon, consequently this carbon has a proportionally strong longing or tendency, so to speak, to contract a further union with oxygen. It is this union of carbon with oxygen, rapidly performed, that constitutes one of the conditions of burning. Hence, alcohol burns very readily indeed.

806. Alcohol mixes readily with water *in any proportion.*

Alcohol may be added to water drop by drop, and it will be found to diffuse itself equally through that liquid, whatever the relative proportions may be. One drop of alcohol may be diffused through a pailful of water, or one drop of water may be mingled with a pailful of alcohol. These two fluids, indeed, agree so well together that it is scarcely possible to obtain them asunder. What is called *absolute alcohol,* the strongest spirit that can be procured, still contains two per cent. of water hidden away in its substance.

807. Strong alcohol *dries moist substances* when they are immersed in it.

That is to say, alcohol has so strong an affection for water, that it abstracts that liquid from most bodies containing it: but the presence of water in dead organic fabrics is essential to their putrefaction. Hence, when they are deprived of their water, they remain unchanged for considerable periods of time. Meat, or any other organized substance, is preserved from decomposition when plunged into strong alcohol, because it is deprived of its water by the superior attraction exercised over that liquid by the spirit.

808. When *alcohol is taken into the stomach* as a portion of the ordinary drink, it is absorbed thence into the blood.

All substances that are perfectly soluble in water, when they are in solution share that pure liquid's privilege of a short and ready entrance into the blood, through the blood-vessels of the stomach, instead of being compelled to go the round about way of the food

(793). Alcohol being itself a thin liquid, and perfectly miscible with water, falls into this predicament. When fermented liquors are admitted into the stomach, both their water and their alcohol pass at once through the coats of the blood-vessels and mingle with the blood.

809. When alcohol is received into the blood, it is carried with its streams *to all parts of the body.*

Thus, whenever alcohol is taken as an ordinary beverage, it in reality becomes a part of the body for the time, for it penetrates into every capillary and fills every crevice with its own peculiar influence, whatever that may be. To drink fermented liquors is, in a degree, to furnish every structure in the frame with alcohol, as well as blood.

810. Alcohol is not *one of the natural and ordinary ingredients* of the body.

None of the more lowly organized animals drink fermented liquors. They all adhere to the use of Nature's simple beverage, water. Man is the only creature that drinks spirit in any form; but there are thousands of the human race who never taste it, and who yet go through long lives, and perform all the offices which men can accomplish, perfectly and well. It is hence clear that alcohol is not one of the necessaries of life: if it were so, Nature would have taken care that it should be supplied to every animated body exactly as water and air are. If alcohol were *essential to animal existence*, it would be found *in every animal frame.*

811. Alcohol is *not nutritious*, and cannot be used by animals as plastic food.

It has been seen that alcohol is entirely composed of carbon, hydrogen, and oxygen. It has no nitrogen in its composition. Hence, it might be inferred that it was not intended to be used for plastic purposes; that is, for building up the structures of the frame (302). But there is another ground for the conclusion that it answers no nutritious object in the economy. It is a substance already withdrawn from the group of organizable principles (227, 228). It is sugar *taken one stage backwards* from organization, towards decay (803). But there is no instance known, where such reduced and simplified compounds take the position of food for animal organization. The law is unvarying and without exception, that it is vegetables which feed upon these simpler inorganic substances, and that animals consume only the complex organizable matters that vegetables build up out of their simple food (252). In almost every case, complex organic matters, when reduced one stage from their organic existence towards the inorganic condition, and in a sort of transition state in downward progress (as sugar is when transmuted into alcohol) prove to be poisons, rather than foods to the animal organization.

812. Alcohol is removed from the blood *by being burnt* **in the circulation.**

Alcohol is more readily combustible than oil, therefore, when it is mingled with the blood, it is at once seized upon by the oxygen that is introduced by the respiratory process, and resolved into water and carbonic acid, which are exhaled from the air cells of the lungs. The four atoms of carbon and the six atoms of hydrogen in each molecule, take twelve more atoms of oxygen than are already present in the molecule, and then the whole of these (4 carbon, 6 hydrogen, and 14 oxygen,) are arranged as four of carbonic acid, and six of water; and thus the process of reduction, which was commenced when the sugar was changed into alcohol and carbonic acid, is finished by the transmutation of the alcohol into carbonic acid and water.

813. When alcohol is contained in the blood, *it is burnt in preference to the oily matters* **naturally supplied as fuel.**

Alcohol is one of the most readily inflammable substances known. If a piece of paper be dipped in spirits of wine and be set light to, the spirits will take fire, but the paper will not be touched by the flame until they are all burnt off, when it too will be consumed. The oxygen of the air has so strong an affection for the combustible elements of the alcohol, that all of it that can get to the flame will combine with them exclusively until they are gone. The alcohol thus saves the paper from burning by its own superior inflammability. The cotton wick of a spirit lamp is never blackened until the spirit is exhausted, for this same reason. The wick of an oil lamp, on the other hand, is charred and burnt as soon as the lamp is lit. Precisely the same thing, as the burning of the spirit in preference to the wick, or to the paper saturated with it, happens when oil and alcohol are mingled in the blood. The alcohol is taken by the oxygen, and burned, and the oil is left.

814. Alcohol consequently prevents *waste combustible matters* from being removed from the blood so freely as they ought (744).

A portion of the fuel, that is burned off in the blood in the production of warmth, is really waste and noxious carbonaceous matter, that needs to be got rid of out of the system. If therefore alcohol is burned, and it is left, the blood is kept impure in consequence. The keeping of the blood charged with alcohol acts somewhat in the same way as the habitual breathing of impure air (747).

815. *Heat is produced* in the body, when *alcohol is burned* in the blood.

Whenever more complex molecules are resolved or split into simpler compounds, heat is of necessity set free and rendered sensible (47, 715). Spirituous liquors are commonly taken freely when men are about to experience great exposure to cold. Under such circumstances, if the blood is kept moving very vigorously by exercise, and the air that is breathed is very rich in oxygen, on account of the compression and condensation of its substance by cold, all the natural fuel of the blood, and all the alcohol mingled with it may both be consumed, and so the body be made warmer in consequence. It is well known that excess in wine or spirit-drinking does less harm in open cold air, than when met in close warm rooms. In the former case, the injurious influence is much more rapidly removed from the blood by being burned, and exhaled from the lungs as carbonic acid and water. But where a provision of this kind is to be made, the taking of oily food, of the nature of fat and butter, answers the purpose far better than the employment of fermented drinks ; for oil produces much more heat than alcohol, when it is burned, although it is kindled into flame so much more slowly. One pound of fat or oil really gives out more heat when burned, than two pounds and a half of brandy. The sense of warmth, that follows upon drinking spirits or wine, is not due to their own heating powers when burned, but to the increased rapidity of circulation, which their stimulant presence in the blood produces.

816. When the blood is kept charged with alcohol, this principle acts, at first, *as a powerful excitement* to most of the vital organs.

Alcohol is an unnatural superfluous ingredient of the blood, and is not wanted there ; hence, whenever it is introduced, Nature hastens to get rid of the noxious intruder as rapidly as she can. She does this by resolving it into carbonic acid and water, and by then pouring these out through the lungs. It is perfectly wonderful how rapidly alcohol is removed from the system in this way. Men may drink wine slowly in the open air, so that the alcohol is poured out of the blood through the lungs as rapidly as it is poured in through the stomach. When, however, the alcohol is introduced more rapidly than it can be got rid of, the blood becomes more and more charged with it, and then the alcoholized blood tells upon every part of the frame. At first it stimulates each organ to increased activity. It makes all work harder to get rid of a presence that is injurious to their healthy existence. This is why upon drinking fermented liquors, various signs of excitement are produced. The heart begins to beat more quickly and more strongly: the skin grows hot, and exhales abundance of perspiration; the secreting organs pour out more of their ordinary productions than they usually do: the features grow flushed; the eyes brighten; and the powers of the mind are quickened. These constitute the ordinary effects of the drinking of wine and spirits, and they are attended by sensations so pleasurable on the whole, that they act as a powerful temptation to men to abandon the pure drink of nature for fermented beverages.

817. The *intellectual powers are deranged* under the stimulant influence of alcohol, *before any of the more material functions* of the body are much interfered with.

It might be expected that the presence of so subtile a mischief in the blood, as alcohol, would first interfere with the perfect working of the most delicate parts of the frame: this really happens. The cerebral masses of the brain are of more exquisite organization, and are more freely supplied with blood, than any other part of the body; hence, if the blood be kept charged with alcohol, the quickened thought that is at first produced, is changed into confusion. The intellectual operations, that are carried on through the instrumentality of the cerebrum, are disturbed. Ideas flow very freely, and gain expression in words, but those words now become foreign to the purpose, and follow each other rapidly and incoherently. The highest faculties of the mind, those of intellect and will, become suspended, even while faculties a degree lower are only roused and excited. Confusion of thought marks the first stage of the influence of alcohol on the nervous organization. The state which is thus induced is appropriately called "*intoxication.*" (The word is derived from the Latin "*in,*" and "*toxicum,*" which signifies "*venom,*" or "*poison.*") When the state of confusion is induced by drinking spirituous liquors, the delicate brain, the seat of reason, is really envenomed and poisoned for the time, so as to be unfit for the performance of its ordinary work.) Alcohol attaches itself to brain substance with peculiar avidity. Animals have had a quantity of spirit poured down their throats, and have then been killed soon afterwards, in order that the effect may be examined, and it has been found that there has been considerably more alcohol in their brains, than in any other portion of their frames of equal size. Now, it will be remembered, that alcohol attracts water from moist substances, and prevents them from undergoing internal molecular change (807); but the brain performs its offices through rapid change in its substance (528). It is in this way, then, that alcohol poisons this important organ, and deranges its powers of healthy action. It takes to itself the water that ought to form a part of the vesicular brain-substance. It shrivels and dries that delicate texture ; but it also appropriates to its own use the oxygen brought by the blood, that ought to be employed in keeping up the vital activities of the part.

818. A state of *languor and depression follows the excitement* that is experienced in the early stages of intoxication, from the use of alcoholic liquors.

If so soon as confusion of the intellect has been produced by drinking fermented liquors, no more be taken, the combustible poison is slowly burnt up in the blood, and so got rid of. That portion which had attached itself to the brain substance is then again taken into the blood, and in its turn disposed of. When the alcohol has been entirely removed from the system, the various organs that had been excited and irritated by its presence, sink into a depressed and inactive condition ; this must be the case, because all excitement of living structures is merely work taken out of them by anticipation, and they sooner or later get exhausted by the undue effort demanded of them (775). The heart's action,

P.

which was strong and quick, becomes weak and slow; the secreting organs cease to separate their ordinary products from the blood; the skin that was moist and warm, gets to be dry and cold; the mind passes from confusion into torpor and lethargy; heavy sleep follows; the over-worked organs get re-invigorated and refreshed by it, and then after a few hours the usual condition of of the body is restored. States of slight intoxication are only followed by temporary discomfort, because nature has established her own means for removing the mischievous influence quickly out of the system, and repairing the injury that has been done.

819. In the *second stage* of intoxication from the employment of spirituous drinks, *the intellectual powers are entirely destroyed* for the time.

If after the mind has been excited, and the ideas and thoughts have been confused, still more of the intoxicating liquor is taken into the stomach, the mischief produced in the nervous organization becomes more extensive and severe. The *cerebrum* soon loses all power of working. Reason and voluntary control disappear entirely, and the instincts and passions usurp the rule from which the reason and the will have been deposed. The *cerebrum* is put entirely to sleep, but the *sensorium* remains still active. In this stage of intoxication, the rational creature, then, has laid aside his reason, and has *sunk himself to the condition of the brute*, being entirely driven by his instincts, and unable to exercise any control over his impulses..

820. In the *third stage* of intoxication, *the sensorial powers* are suspended, and placed in abeyance.

If men go on drinking after they have lost their reason, the blood gets so much more charged with the deadening poison, that at length the less delicate sensorium becomes paralyzed as well as the cerebrum, and

then the loss of sensation and consciousness follows that
of intelligence. First, giddiness is experienced; then
strange sounds are heard, odd sights seen, and a thousand illusions produced. Next the eyes become vacant,
the face grows pale, the muscles tremble and lose their
power, the limbs refuse to perform their offices, and at
length the drunkard falls into his appropriate position,
and grovels in the dust.

821. Men generally *recover from the insensible state* of intoxication, because they are no
longer able to swallow more of the intoxicating poison.

Nature has beneficently provided that when men have
all but killed themselves by swallowing poisonous
alcohol, an end shall be put to their folly by inability
any longer to perform the mere mechanical effort of
carrying more drink to their mouths. So soon as they
cease to supply the alcohol to the stomach, it begins to
be burned out of the blood, and cleared off through the
lungs. In a few hours, therefore, the sensorium resumes
its activity, and they become sensible again. Next the
cerebrum reawakens to its powers, and first, confused
thought, and then clear notions arise. After this, sleep
comes to repair the exhaustion.

822. Drunkards *do not always recover* from
the insensible stage of intoxication.

It sometimes happens that so much alcohol has been
received into the stomach before insensibility sets in,
that the blood is kept fully charged with it, notwithstanding its removal through the lungs, and its effect is
then felt in the spinal cord, as well as in the sensorium
and cerebrum. So soon as the spinal cord has its structure so influenced by the presence of alcohol, that it can
no longer perform its offices, all the various muscular
movements, which it ordinarily keeps up, those of respiration in the number (554) stop, and suffocation consequently is produced. It was intended that the sen-

sorium and the cerebrum should sleep sometimes, consequently life is not immediately destroyed when their sleep is the torpor of intoxication; but it was also designed that the spinal cord should never sleep (555). When it is rendered torpid by intoxication, life is immediately extinguished.

823. The *fourth stage* of intoxication *is* *death.*

Whether a man recovers from the insensibility of intoxication or not, depends upon the accident of his having swallowed a few drops more or less of the poison, under the load of which all his higher vital privileges are crushed for the time. No one who has acquired the vice of drunken habits ought ever to deceive himself upon this point. When he is insensible from drink, he is in the last stage but one of poisoning. Unconscious and unable to lift his finger to help himself, he is lying over the brink of the fearful abyss, from which there is no return for the fallen. He has placed himself there by his own perverse folly, and there he must remain until accident decides whether he is to be cast forth into its fathomless depths, or to be withheld yet for a little season to meet other pains and penalties, before the end of all arrives. In his existing state, it may be only the weight of a hair that is to determine for him between immediate death and prolongation of his life. It only needs that a little more alcohol should be accumulated in the blood, and the spinal cord will be rendered inactive under its stupefying presence, as well as the sensory and intellectual organs, and then the play of the chest, which is kept up by its influence, will be stilled; respiration will cease; venous blood will be sent in addition to the alcohol, to where arterial blood ought to flow; and a few failing throbs of the heart will end the life that has been prized so lightly, and thrown away so guiltily.

There are thus four distinct stages of intoxication, and the one passes into the other, as more of the alcoholic poison is accumulated in the blood.

The first stage principally affects the heart, and is characterized by general excitement.

The second stage principally affects the cerebral masses of the brain, and is marked by the loss of reason. *This is the stage which makes man a beast.*

The third stage affects the sensorial masses of the brain as well as the cerebral, and is marked by loss of sensibility. *This is the stage which sinks man below the level of the beast.*

The fourth stage affects the spinal cord as well as the brain, and is characterized by death. *This is the stage that makes man a corpse.*

824. Frequent intoxication *destroys life in a slower way*, even when it does not put an end to it at once.

The majority of persons who indulge in habits of drunkenness, do recover from the "*beastly*" state time after time, on account of the facts alluded to already, namely, that so soon as they are insensible, they cannot go on drinking, and that nature then carries on her work of burning and dissipating the poisonous agent. Still they have to pay a very severe penalty, nevertheless. Death is only eluded for a little time. The sure avenger is upon the track of his victim. The repeated derangement induced in the actions of the various vital organs, saps the foundations of vitality, and changes health into disease. A sober and temperate man, at the age of twenty years, has the fair prospect of living for forty-four more years; but it has been well ascertained that a confirmed drunkard, at the same age, can reasonably look for only fifteen more years! The drunkard therefore pays the penalty of twenty-nine valuable years entirely sacrificed to his vice, besides the additional time that he gives to the insensibility and stupor of his drunken fits.

825. Habits of intoxication *induce disease* by rendering the blood unfit for its purposes,

and by debilitating and exhausting the various vital organs.

When alcohol is kept almost constantly present in the blood, that liquid is rendered less plastic than it ought to be (361), so that loose and flabby fibrinous structures are built up from it instead of firm ones. The poison, too, occupies the place some more natural ingredient ought to hold, and in this sense impoverishes the fluid it is mingled with. Besides this, it takes the oxygen that ought to be consumed in the removal of waste material, for its own combustion (814), and so keeps the blood laden with noxious impurities, that nature designed should be removed as soon as they were produced. Alcohol also first excites, then exhausts, then irritates, and then inflames and destroys the structure of the numerous organs through whose capillary vessels it passes, when circulating with the blood. Nothing so much weakens the resisting powers of the frame, and disposes it to suffer grave disorder, when other disturbing influences are brought into play (780), as the frequent employment of intoxicating drinks.

826. The *moderate use* of alcoholic liquors produces much slighter effects than those that are induced by the immoderate employment of the same agents.

When fermented liquors are used in moderation as the ordinary beverage, the alcohol that is introduced by them into the blood is so dilute, and is in such small quantity, that it is generally expelled from the system, almost as soon as it is received, and it never produces any other result than general stimulation. Its bad effects are then so slight, that they may be encountered for years, and yet nature be able to find some way of compensating for them and making all things straight. The system is merely kept working a little too high, yet not so high but that it may be able to right itself by prolonged sleep, increased exercise in pure air, the occasional use of medicinal influences, and other like appliances. Still, even in these cases, the alcohol is burned off in preference to impure matters that ought to be removed from the blood. The rich plasticity of that fluid is impaired, and the various vital organs of the frame are kept in alternate states of excitement and depression. In short, even then the risk of disorder is slightly increased in consequence of the artificial indulgence.

827. Persons, who are in the enjoyment of vigorous health, act wisely when they *refrain altogether* from the use of fermented liquors.

It is quite clear that he must be the wisest man who avoids exposing himself even to a slight risk of mischief. It is plain that when alcohol is taken into the blood, it excites the vital organs to exalted activity, and that this exalted activity is followed by consequent depression and exhaustion. It is manifest too, that nature views its presence in the blood as, at least, superfluous, and gets rid of it thence as rapidly as possible. Bearing these facts in mind, and connecting with them the unquestionable truth that myriads of living creatures enjoy uninterrupted vigor and health without ever admitting a single drop of the alcoholic stimulant into their frames, and also the important reflection, that the exhaustion which follows excitement from alcoholic stimulus leads men involuntarily to crave a renewal of the excitement, and so steadily to advance in the evil course when once they have entered upon it,—it must be apparent to any rational being, that water-drinking is safer than spirit-drinking, in even its most dilute and least noxious form. The great reason for the extensive employment of alcoholic beverages among mankind is found in the fact, *that nearly all those who employ them have experienced their stimulant and exhaustive effects before they know what they are doing, and so are insensibly trained to crave renewed excitement from them.* In civilized society, man lives in a state of constant over-excitement. All the arrangements about him tend more or less to exhaust his powers; hence, whatever rouses him for the time, commends itself to his use by the agreeable feelings it causes to take the place of uncomfortable ones, and he looks no further than the present, and yields to temptation. This is why people are generally so fond of· wine and beer. There is, however, danger in this yielding, exactly proportioned to the extent to which it goes. This danger men would never expose themselves to if they really understood what they were doing before they contracted the unnatural craving, and if they really knew the price they of necessity paid for the indulgence.

828. When wine and beer *are* used as ingredients of the ordinary diet, they ought to be taken *in the place of some other article* of food.

Although persons, who are in vigorous health, are best without fermented liquors, these may prove to be valuable aids, (if judiciously taken,) to persons who are not in vigorous health. In other words, fermented liquors seem to have been intended for medicines rather than for food. Both wine and beer, and especially the latter, contain complex plastic principles (that is, composed of all the four organic elements) [122], mingled with their water and alcohol. In beer this plastic principle is the gluten of the barley from which the malt is made. In cases, therefore, in which the body is insufficiently nourished, in consequence of any inherent fault of structure, both plastic matter and fuel may be at once added to the blood, by the employment of these beverages, in just the quantities that are required, without the necessity of making any great demand upon the powers of the digestive organs. But then it must always be remembered, that these beverages take the place of a certain quantity of the ordinary food. It is a very interesting fact, that whenever fermented liquors are productive of service in the economy, instead of leading to mischief, no excitement is experienced on using them. Their stimulant powers are absorbed in remedying depression and exhaustion that are already present, and so, in raising the forces of life to their ordinary standard, instead of being employed in elevating them to a point of danger above this standard. Some very carefully conducted experiments and observations recently made by German physiologists, have shown that the habitual employment of these beverages stops or delays the waste of the fabrics of the body. They have also proved that water-drinking quickens that waste. Now, *as the object of animal life is the support of waste, and not its stoppage* (258), water-drinking must be better than wine and beer-drinking, for vigorous healthy men who desire to get the full amount of work out of themselves of which their frame is capable. The more rapid the waste of structure, the greater the muscular and nervous effort that can be made. When, on the other hand, such results cannot be secured without wearing out the body, it is very desirable to stop the waste, *since it is better to live a half life, than to have no life at all.* It is under such circumstances that the judicious employment of fermented beverages may be made to lengthen life. In all cases of this kind, the unused structures of the frame are consumed by the oxygen that is introduced in respiration, instead of only the products of their decay. Consequently it is then well, if the presence of alcohol in the blood takes the sharp tooth of the consumer—the oxygen—off from them. This is altogether different from the condition in which, on account of rapid natural activity and waste, there are large quantities of worn-out matter constantly in the blood, and requiring removal.

829. Tea, coffee, and cocoa, *are nutritious* and stimulant beverages.

The experiments of the German physiologists, alluded to above, prove that tea, coffee, and cocoa, possess the same power of arresting waste in the structures of the organic frame, as wine and beer. There is, however, this great advantage attending their habitual use, that they contain no alcohol. None of the injurious effects shown to be inseparable from the influence of that agent can be produced in their case. Tea, coffee, and cocoa,

24   P*

really contain a considerable quantity of nutritious plastic matter, like beer (828). This plastic substance may easily be procured from tea in its pure state for examination. If a little tea be placed on a watch-glass, and this be covered by a piece of paper, and then heated upon an iron plate until a brown tinge appears, long, white, glittering crystals will be seen to be adhering to the surface of the paper and to the tea. These crystals are composed of a peculiar compound of carbon, hydrogen, nitrogen, and oxygen, upon which the name of "*theine*" has been conferred. Each molecule of this crystalline principle is formed as expressed in the annexed symbol. Coffee and cocoa contain a similar substance, almost identical in composition. "Theine" and its allies seem to be really plastic food capable of being converted into organized structure. Indeed they deserve to be ranked amongst the organizable bodies (227) formed by vegetables as the food of animals. Tea also contains small quantities of iron in the same highly active state in which the mineral is present in medicinal chalybeate springs. There are reasons for the notion that theine, and the similar plastic principles found in coffee and cocoa, serve particularly as food for the nervous structures. They seem to possess some peculiar power of supporting the operations of these parts of the frame. When employed in excess they undoubtedly produce nervous excitement in an injurious degree.

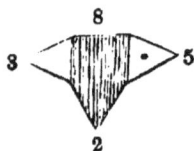

CHAPTER XXVI.

DECAY, DISEASE, AND DEATH.

830. All complex organic bodies *decay* when they are removed from the influence of life.

That is, they fall to pieces, and lose all vestige of organic or complex structure. The nature of decay is well illustrated in what happens to dead leaves in the autumn. As they blow about on the ground, their fibres gradually exhale into vapor and fall into dust, so that by the time the spring returns, their forms have vanished from the face of the earth. The term "*decay*" very well expresses this result. (It is derived from the Latin words "*de*" and "*cado*," which signify "*from*" and "*to fall*." Things that decay fall from their perfection, and crumble away.) "*Decompose*" is another term that is in common use for expressing the same change. It is also taken from the Latin. A thing is said to be "*composed*" when certain elements are "*put together*" ("*cum*" and "*pono*") to form it, and consequently, when those elements are put asunder to *unform* it, it may very appropriately be said to be "*decomposed*."

831. The decay of organic bodies is effected *through the influence of corrosive oxygen* (35).

The peculiar combination in which the several elementary atoms are held in complex molecules, is dis-

solved so soon as those molecules are abandoned to the
influence of the air. The oxygen of that air possesses a
stronger attraction for the various elements that are so
combined together, than they do for each other. Consequently it dissolves the union, and carries off one after
the other until all are gone. Organic bodies invariably
decay when left to natural arrangements, because they
are always within the influence of the atmosphere.
Whenever they are so placed that air cannot get to
them, they do not decay. It is on this account that
organized structures may be preserved for long periods
of time when immersed in closed bottles of alcohol, or
even when soldered into air-tight cases of metal.

**832. When organic bodies decay, their
substance is finally *resolved into carbonic
acid, ammonia,* and *water.***

That is to say, they are again converted into the
materials from which they were primarily formed. For
it will be remembered that these three compounds are
the principal food of plants (125), and that plants make
organizable substances out of their food (124). The
presence of air is essential to the resolution of complex
organic matter into the simpler compounds named
above, because none of them contain so much oxygen
proportionally as the equivalent amount of carbonic acid
and water. Thus for instance, every molecule of sugar
possesses 12 atoms of carbon, 12 of hydrogen, and 12
of oxygen (228); but the carbonic acid and water, into
which it is resolved by decay, consist of 12 atoms of
carbon, 12 of hydrogen, and 36 of oxygen. The 12 of
hydrogen want all the 12 of oxygen for their conversion
into water (75); consequently the 12 of carbon must
find the 24 of oxygen needed for their transformation
into carbonic acid (105) elsewhere. They in reality
procure them from the store kept ready in the atmosphere (59). The carbonic acid, water, and ammonia,
formed when organic bodies are decomposed, fly away
as invisible vapor into the air. As they do so, the

small quantities of subordinate mineral ingredients, with which they have been combined (126–128) being of a fixed nature, are returned as ashes to the soil (129).

833. But all complex organic bodies also decay *even while they are under the influence* of animal life.

There is really no difference, *in this respect*, between the fate that is encountered by organic bodies while contained within the living frame, and after they cease to form a part of its arrangements. For, in the former case, they are not removed from the influence of the corrosive oxygen. Streams of it are continually conveyed to them through the air cells of the lungs, and by the instrumentality of the blood corpuscles (411, 425). The only distinction between the two conditions is, that while in the one, new organic matter is supplied to the structures as fast as the old is consumed by decay, in the other, no such renovation is effected. So long as organs are alive, their textures and bulk are maintained entire, even while the substance, of which those textures are made, is undergoing continued consumption (265).

834. When the substance of living structures *decays within the body*, it is resolved into carbonic acid, water, and ammonia.

The complex substances, of which the various vital organs of the frame are built, are being constantly consumed under the influence of the oxygen that is conveyed to them through the blood channels, and are being changed by the consumption into carbonic acid, water, and ammonia, exactly as occurs with dead organic matter, abandoned to the air without the precincts of the living frame.

835. Carbonic acid, water, and ammonia are therefore *the principal waste matters* that have to be removed from the living frame, as the products of its destructive activity.

Carbonic acid, water, and ammonia are exhaled from all living animals, and thrown into the atmosphere by which they are surrounded. It is in this way that animals, which are fed by plants, return the service by furnishing them with constant supplies of their food (257).

836. The removal out of the body of the waste matters formed by the decay of the substance of organs *is termed excretion.*

"*Excretion*" is derived from the Latin words "*ex*" and "*cerno*," (which signify "*from*" and "*to separate.*") The worn-out substance of the various structures of the frame is at first mingled with the circulating blood in the capillaries. It is then conveyed to organs especially fitted for separating it from the blood. The organs that effect this purification are termed excreting organs.

837. *The liver* is an excreting organ.

The liver is a large, fleshy organ, packed away in the abdomen, immediately above the stomach (299). This organ has been already spoken of, as rendering very important aid in the work of digestion (291). It, however, performs other service also, which is equally essential in the animal economy, and which must now be noticed.

838. All the veins that *carry blood from the digestive organs* run to the liver.

The veins that run from the different organs of the abdomen, are all united into one large venous trunk, which plunges into the fleshy substance of the liver, instead of at once continuing its course towards the heart. This large duct, that carries venous blood to the liver, is called the "*portal vein.*" (The word is derived from the Latin "*porta*," a "*gate*," or "*entrance*.") It was so termed in the first instance by the old anatomists, who saw it pass into the large organ between pillars, as if through a gate.

839. The venous trunk, that runs to the

liver, *subdivides within its substance* into a great number of capillary channels.

The portal vein branches out in the liver into minute divisions, just as the great vessels, that issue from the right side of the heart, branch out into capillary divisions in the lungs (420).

840. The capillary vessels of the liver *lead to larger and larger venous channels*, that at last end in two veins, which are called *the hepatic veins*.

The capillaries of the liver thus, like the capillaries of the lungs, and other parts of the body, end in veins (401) which receive their blood from them as it moves onwards. These veins are called "*hepatic veins*," (the word being derived from "*hepar*," the Latin for "*liver*.") Thus the main peculiarity of the liver is that *great veins branch out into it after the manner of arteries,* and are then again gathered up into veins. Venous blood flows by a system of ramifying channels to all parts of the liver, and then passes onwards through it to be collected in two common channels. The veins that distribute venous blood to the liver are the " veins of the gate," or "*portal veins*." Those which receive it after it has passed through the liver, and convey it away, are the proper veins of the liver, or " *hepatic veins*."

841. The *chief mass of the liver* is formed of a great quantity of living cells.

These living cells are placed in the midst of the capillary terminations of the portal veins, and are abundantly supplied with blood from them.

842. The cells of the liver *separate bile* from the venous blood (291).

The liver in fact is a large secreting organ (447, 448). Its living vesicles gorge themselves with the peculiar liquid known as bile, taking the materials for its formation from the blood.

843. The bile, that is formed by the liver-cells, is *received into a series of tubes* provided for the purpose.

A quantity of fine tubes run in the midst of the capillary veins and living vesicles of the liver; these are called the "*hepatic ducts.*" As soon as the vesicles have gorged themselves with bile, they fall off into these ducts, carrying their contents with them (448, *Figure* 42).

844. The hepatic ducts of the liver *terminate in one main channel*, which finally opens into the bowel (291).

This main channel, into which all the smaller hepatic ducts are finally collected is called the "*bile duct.*" The bile duct carries the bile, formed in the liver, to the bowel, and empties it into that cavity.

845. The hepatic veins end in one of the great vessels that *carry venous blood to the right side* of the heart (418).

The liver thus deprives venous blood of bile, and then sends it on to the heart, to be pushed forwards into the lungs. The liver is a purifying organ. The purification that it effects is the removal of bile from the blood.

846. Bile is an oily substance *converted into a sort of soap* through the influence of soda.

Soap is merely oil mingled with potash or soda. These compounds possess the power of making oil soluble in water (which it is not naturally), when added to it. Oil, when rendered soluble by admixture with potash or soda, is converted into the peculiar substance known as soap. There is always soda in the blood (mostly derived from the oxidation of common salt taken with the food). This soda is employed in the formation of bile.

847. The oily ingredient of bile *is derived from the waste of the highly carbonized textures* of the body.

All oily bodies are mainly composed of carbon and hydrogen (235). Bile consists of carbon, hydrogen, and soda. Its carbon and hydrogen are the carbon and hydrogen that are thrown as waste into the blood, upon the destruction or decomposition of textures containing those elements. The waste carbon, and a considerable portion of the waste hydrogen of the body, are separated by the liver from the returning venous blood, and are then thrown out into the bowel in the condition of bile.

848. The bile that is poured into the intestine, *is again taken up from it into the blood.*

After the bile has performed certain useful offices in the digestive process (292), it is drunk up through the walls of the bowel, and so introduced into the blood-vessels, and mingled with the circulating fluid. It is composed, it is true, of refuse matter that needs to be removed altogether out of the system; but this refuse matter has still one kind of virtue in it, and this virtue is taken advantage of, before it is finally thrust out of the frame.　·

849. The bile *is burnt as fuel*, after it has been mingled with the blood.

That is to say, the reduction of the worn-out substance that was commenced by resolving its carbon and hydrogen into bile, is completed in the blood, under the influence of the oxygen mingled with it there, by the entire change of the carbon and hydrogen of the bile into carbonic acid and water. Bile is a very inflammable substance, and yields heat when it is resolved into carbonic acid and water. Hence, it is good for fuel when it is good for nothing else. On this account, it is not at once thrown out of the system, when its elements are removed from the various organs in which they have been doing duty, but is employed in the heating service of the body, and so burnt off instead. The waste carbonaceous matters of the body are changed into bile before they are employed as fuel, for two reasons. In the first place, they are more readily combustible in this state than when in their more rude condition : and, in the second place, bile is comparatively harmless in the blood, although the waste matters from which it is formed are absolutely poisonous when retained in the fluid. The liver takes noxious products out of the blood, and makes them bland, in order that they may be then returned into the circulating fluid when no longer dangerous, and be there employed as fuel. The change of waste and noxious carbonaceous matters into harmless fuel is thus a piece of very interesting economy. When more fuel is required for the preservation of the animal temperature than the waste of the carbonaceous principles of the body can supply, portions of the food that are of analogous nature, are at once mingled with the soapy mass. Sugar and some kinds of fat are served in this way.

850. The carbonic acid and water that are formed by the burning of the bile in the blood, *are exhaled through the air cells of the lungs.*

In this way, then, the carbon and a part of the hydrogen resulting from the waste of the textures are finally expelled from the system. The carbonic acid and some of the water formed by the decay going on within the living system (834), escape through the lungs to mingle with the external atmosphere, and to become in it the food of vegetable life. The liver and the lungs are both excreting organs (836). They are outlets through which portions of the decayed parts of the frame are being continually poured away. The liver, it will be remarked, is really an auxiliary piece of apparatus to the lungs. It prepares some of the matter that is removed through the lungs, to be easily and safely exhaled from their air cells. It is very remarkable how really alike these orgaus, apparently so distinct, are in their natures. They are both supplied by venous blood, and both separate from it carbonaceous matter. The only real distinction is, that the separated matter is liquid in the case of the liver, and is, therefore, collected in ducts and conveyed away by them, while it is a vapor in the case of the lungs, and is steamed out at once through the walls of the vessels. The venous capillaries of the liver run amongst vesicles, which can absorb the liquid bile from them by the operation of osmose (163); but the venous capillaries of the lungs merely run amidst air cavities into which they can steam out their exhalation.

851. The ammonia that is formed from the waste of the textures of the living body *cannot be exhaled in its gaseous state.*
It will be remembered that ammoniacal gas is a very acrid, pungent substance (114). On this account it cannot be allowed to come in contact with any of the living textures in its fully formed state. If, for instance, ammoniacal gas were exhaled with the carbonic acid gas through the air cells of the lungs, it would produce most painful irritation in them. Carbonic acid is free from all acridity or pungency, and therefore can be produced within the body, and can be immediately thrown out through its living outlets. But ammonia must be dealt with in another way.

852. The ammonia produced by the waste of the living structures of the body, *is carried out from it as a peculiar substance,* to which chemists have given the name of *urea.*
This urea, like bile, is contained within the blood; but it is formed in that liquid without the intervention of any organ. So soon as the elements of which it consists are thrown together in the capillary vessels, on account of the decay of the textures of the body, they at once combine to form it.

853. Urea contains within itself *all the four organic elements* (122).

Ammonia consists, it will be remembered, of hydrogen and nitrogen (116); but so long as they continue within the living vessels, small quantities of carbon and oxygen are kept attached to them to suspend the formation of the acrid ammonia. Just as the decaying carbonaceous matters of the frame are kept in the blood for a time as soapy bile, before they are ultimately resolved into carbonic acid, the decaying plastic (nitrogenized) matters (302) are kept for a time in the state of bland urea before they are finally resolved into ammonia. Each molecule of urea has a composition that may be symbolically expressed, as in the annexed diagram.

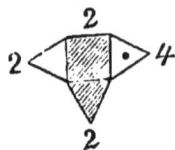

854. Urea is separated from the blood *by the kidneys.*

The kidneys are thus excreting organs, as well as the liver and lungs; but they are designed for the removal of the waste plastic matters, such as albumen, fibrin, and hæmatin, out of the system. The kidneys are placed in the cavity of the abdomen (299) behind the liver, and in front of the backbone, one on each side. Each kidney consists of an arterial branch subdividing into capillaries, which lead on into veins in the ordinary way. These capillaries are meshed about delicate tubes, that are lined by secreting cells, and that are all gathered together at last into one duct, which conveys away the separated urea out of the body. The kidneys thus are of similar structure with the liver.

855. Urea is *very soluble* in water.

Hence, it is always removed from the blood in solution, and is poured out from the body in a liquid state. It is devoid of all pungency, until it begins to be changed into ammonia by further decay, and is not at all acrid. It is more like a salt than any thing else. Nearly an ounce of it is produced by the decay of albuminous matters, and got rid of, every twenty-four hours.

856. Urea is *changed into ammonia and carbonic acid*, after it has been thrown out of the body.

That is to say, the decay of the albuminous substance is then completed. It will be observed that it is only necessary to add two compound atoms of water (75) to each molecule of urea (853), in order entirely to resolve it into carbonic acid and ammonia, leaving nothing behind. Thus a molecule of urea consists of

	Carbon.	Oxygen.	Nitrogen.	Hydrogen.
	2	2	2	4
Add to these 2 atoms of water........ $\Big\}=$		2		2
and the result is........	2	4	2	6

but 2 carbon + 4 oxygen = 2 carbonic acid; and the 2 nitrogen + 6 hydrogen = 2 ammonia.

Thus, then, the textures of the living body are changed by decay into carbonic acid, water, and ammonia, exactly as organic matters out of the body are (832), and these compounds are thrown out by the liver, lungs, and kidneys. The carbonic acid principally, with watery vapor, through the liver and the lungs; the ammonia and some of the carbonic acid, as urea, in solution in water, through the kidneys. The liver, the lungs, and the kidneys are therefore the great excreting outlets, through which the waste of the frame is being continually poured away.

857. The *fixed and subordinate ingredients* of the decaying textures of the body (127) *are removed from it with the urea*, in solution in water.

It will be remembered that besides the four organic elements, there are small quantities of other more fixed ingredients contained in organic textures: as, for instance,

potash, soda, phosphorus, sulphur, iron, lime and salt
(127). When the various textures that contain these
elements, decay, or are worn out, small quantities of
them are thrown into the blood in the condition of soluble salts, and are then carried away in solution, with the
urea. They are mostly in the state of neutral *sulphates*,
or *phosphates* (149). For the sulphur and phosphorus
get converted into sulphuric and phosphoric acids (132),
by the influence of the oxygen circulating in the blood,
and these acids then combine with other substances of
the oxide series (149) to form ternary compounds. It
is a curious and very interesting fact, that, whenever
any prolonged mental application is made, a proportionally increased amount of phosphate salts is thrown
out of the body in solution in this way. For the brain
substance contains a great deal of phosphorus (525),
and, therefore, more phosphorus is set free, when more
of it is wasted. Great and long-continued muscular exertion, on the other hand, leads to an augmented excretion of urea, for the muscles contain a great deal of
nitrogen in their composition.

858. Decaying organic matters *are thrown
out from the body also* through the air cells
of the lungs (724), through the pores of the
skin (781), and through the coats of the
bowels.

The impure exhalation that is thrown out with the
watery vapor of the breath (724), consists of albuminous and other plastic matters in a state of commencing
decay. The lungs, therefore, help the kidneys to a certain degree, in the removal of ammoniacal matters, besides effecting their own particular work of exhaling
carbon. The solid ingredient of the perspiration (781),
and also certain excretions of a similar nature within
the bowel, are of a like character. There are, therefore, five different structures that are mainly concerned
in effecting the removal of waste matters out of the
system — that are in fact *its excreting organs:* these

are *the liver*, *the lungs*, *the kidneys*, *the skin*, and *the bowels*.

859. When plastic substances of a very complex nature decay, their elements generally *pass through intermediate states*, before the final resolution into carbonic acid, water, ammonia, and mineral salts, is effected.

That is to say, in such complex bodies, there are so many elements present, that when the first union is loosened, other temporary combinations, still of a somewhat complex character, are almost sure to be formed amongst them. Alcohol is one illustration of this intermediate stage in the downward progress. When the union which constitutes a molecule of sugar is dissolved, the various elements form molecules of alcohol and atoms of carbonic acid, before they are entirely changed into carbonic acid and water (802). The formation of bile (847) and urea (852) in the system, out of the refuse material of the textures, is another instance of this step-by-step progress downwards in decay.

860. When very complex organic substances decay, the process is generally termed *putrefaction*.

Putrefaction is simply distinguished from decay in the fact that during its progress, intermediate compounds possessing disagreeable odors, are formed. The word "*putrefaction*" is derived from the Latin terms "*putris*" and "*fio*," (which signify "*putrid*," and "*to become*"). Bodies that consist exclusively of carbon, hydrogen, and oxygen, decay: that is, they are changed at once into carbonic acid and watery vapor, both of which are devoid of smell. This is seen in the burning of charcoal, coal, and oil, and in the fermentation of sugar; but bodies that also contain nitrogen, phosphorus, and sulphur, besides the carbon, hydrogen, and oxygen, putrefy: that is, they are changed into various complex substances, which are volatile, and possess unpleasant odors, before they are ultimately resolved into carbonic acid, water, ammonia, and phosphates or sulphates. Phosphorus and sulphur unite with hydrogen, in particular, to constitute some peculiarly disagreeable compounds. The impure exhalations of the lungs, skin, and bowels, are of this nature.

861. Most of the intermediate compounds that are formed during the putrefaction of very complex organic substances are of a *poisonous nature.*

One of the simplest of these intermediate substances is alcohol, which is really a product of decay, rather than of putrefaction, being formed out of a three-element substance. Yet it has been seen that even this is a very powerful poison, when received into the system in large quantities (817). The much dreaded and highly energetic poisons, *prussic acid* and *oxalic acid*, are intermediate compounds, formed by complex bodies in their progress towards destruction, and serve very well to illustrate the point under consideration. The molecule of prussic acid has the composition expressed by the annexed symbol. It is formed by the decay of a complex albuminous principle made by certain plants of the almond and cherry tribe. It will be observed, that by the addition of two atoms of water, and two of pure oxygen, the molecule of prussic acid may have its resolution completed, and be converted into one atom of ammonia, and two of carbonic acid. Oxalic acid is sometimes produced even within the animal body, by the decay of certain of its complex principles, but is then immediately thrown out with the urea. The composition of the molecule of oxalic acid, is symbolically expressed by the annexed diagram. It will be observed, that the molecule of oxalic acid only needs to have one other atom of oxygen added to it to be resolved into two atoms of carbonic acid.

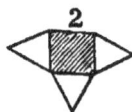

862. These intermediate products of decomposition are poisonous to living animals, on account of their exercising *very powerful chemical and physical influences* over the textures of their frames.

Most of these half-organic poisons exercise so powerful an influence over the still living structures of the animal frame, that they derange the ordinary relations and operations of their elements. It has been seen that alcohol so interferes with the perfection of the brain structure, when placed in contact with it, that it entirely arrests its usual activity. Prussic acid is far more energetic than alcohol. A single drop of it introduced into the mouth, or a full charge of its vapor in the lungs, will cause the strongest man to drop down dead in a moment.

863. Some of these half-organic poisons injure the living structure they come in contact with, by inducing it *to resolve itself into a similar decomposing state* with themselves.

They do for the living structure what yeast does for the gluten that is contained in a solution of malt when it is mingled with it. They cause a sort of fermentation or decay in it; and then very commonly this fermentation or decay is attended with the production of fresh quantities of the half-organic poison out of the decaying substance, just as the fermentation of malt liquor, through the agency of yeast, is attended by the production of fresh quantities of yeast.

864. The impure exhalation *thrown out from the blood in the breath* (724) *is a half-organic poison*, produced by an early stage of decay.

It is produced by the decay of some of the complex plastic and albuminous principles of the blood, or other parts of the body. It is thrown out from the blood

25   Q

because it is poisonous while retained there. If it is breathed over and over again in close rooms, it produces most grievous results, and even death. One of the objects aimed at, in making arrangements for the ventilation of close apartments, is the removal and destruction of this poisonous exhalation from the human lungs (742).

865. In ordinary circumstances, the poisonous exhalation of the animal breath is *destroyed and rendered harmless* by free admixture with pure air.

The air effects this beneficent purpose in two ways. In the first place, it dilutes the exhalation so much under the operation of the law of gaseous diffusion (58), that its peculiar energies are at once diminished many-fold: and in the second place, the free oxygen of the air almost immediately carries the decay still further, and completes the reduction of the dangerous principle into water, carbonic acid, and other like inert compounds. Hence, when the air that is breathed is kept tolerably fresh, this poisonous exhalation of the breath is put out of the way as soon as it is thrown forth from the mouth, and effects no harm.

866. Under peculiar circumstances, the impure exhalation of the animal breath *becomes more than usually powerful* and energetic for mischief.

It has been stated that this happens when it is thrown into confined spaces, in which the air is not duly changed, and in which consequently there is a deficiency of oxygen for perfecting its resolution. It then accumulates more and more, until it becomes unusually energetic, merely on account of concentration (*that is, augmented quantity in any given space*). Sometimes, however, it is really thrown out from the lungs, already in a condition of augmented virulence, so that a very slight concentration indeed in it is sufficient to lead to the worst results.

867. The exhalation of the breath becomes

unusually poisonous and energetic for mischief, whenever *the blood within the body is
made extraordinarily impure.*

The blood still circulating within the body may be
rendered impure in two several ways. The products of
decay, that are constantly thrown into it through the
destructive operation of the living structures, may not
be as quickly and efficiently removed through the excreting organs (836); or, impure exhalations may be
floated on the air into the air cells of the lungs, and so
may be communicated directly to the blood lying in
contact with their membranes (748). Under either circumstance, or when both combine, the blood will of
necessity become heavily laden with impurities.

868. *Fevers* are conditions in which *the
blood is in a very impure state.*

The class of disorders characterized as fevers are produced in both of the ways indicated above. By some
accidental circumstance, the organs entrusted with the
due removal of the excretions are for a time rendered
incapable of performing their proper work; or very impure air, loaded with poisons of a complex, half organic
kind (861), is breathed for some time: in either case the
blood gets very impure; but this impure blood being
carried to every organ of the frame, in the due course
of circulation, disturbs its ordinary operations. Each
organ then endeavors to get rid of the unusual and
offending matter that is brought to it, by working
harder; that is to say, it labors, instead of transacting
its business quietly and well. Hence, the breathing is
quickened; the heart beats irregularly and hurriedly;
the skin becomes chilly and hot by alternate fits; the
appetite fails, and the stomach is attacked with nausea;
the body and the mind are both oppressed with weariness; the sleep is unrefreshing and broken by disturbing
dreams, or it takes to flight altogether; aching and pains
of various kinds are experienced in the limbs and different parts of the body; the several secretions are dried

up, or they become very depraved; and, in short, the
usual state of comfort and ease is exchanged for a condition of very disagreeable struggle and conflict. These
are the symptoms of the disorder to which the physicians
have given the appropriate name of "*fever*," on account
of the ordinarily hot state of the skin that prevails during
its continuance. (The term is derived from the Latin "*fer
veo*," to be hot.) Fever is simply an oppressed state of
the system, in which every organ comprised within it is
supplied with impure blood, and is engaged in the labor-
ious effort to get rid of the offending matter that is
brought into its channels.

869. Some fevers are of an *infectious nature.*

That is to say, they are capable of communicating a
precisely similar state of disorder to apparently healthy
people brought within their sphere of influence. The
word "*infection*" is derived from the Latin "*inficio*"
(which means to "*contaminate*," or "*affect with the same
quality*"). Infectious fevers affect persons previously
free from obvious disorder with their own contamina-
tions.

870. Infection is a *half-organic poison of
a very subtile nature,* formed by the early stage
of decay of a complex organic substance
(861).

It is indeed merely the ordinary impure exhalation of
the breath, rendered unusually energetic for mischief, in
consequence of the impure state of the blood (867).
When the blood is kept thus impure, concentrated exhalations are poured out from putrefying matters contained in it, which are more mischievous (so far as their
influence on living structure is concerned) than those
which are commonly formed. These exhalations are,
however, of so subtile a nature, that they cannot be detected by direct observation of the senses. Sometimes
they are perceptible to the organ of smelling, but some
of the most deadly of them give no other sign of their

presence, than such as is afforded by the mischief they work.

871. The poison of infection is of somewhat the same nature *as a ferment*.

That is to say, it is able to produce a subtile half-organic poison precisely like to itself, from the blood that it mingles with, exactly as yeast produces fresh yeast in fermenting malt liquor. It is in this way, then, that infectious diseases are communicated from person to person. A subtile poisonous exhalation is breathed, which produces in the blood more subtile poisonous exhalation like to itself. This is then poured from the lungs into the air, and is in turn breathed by other people, mingles with their blood, produces there more subtile poison, and so the infection spreads.

872. Disorders, produced by the ferment-like action upon the blood of an inhaled poisonous exhalation, are called "*xymotic diseases*."

Hence, all infectious disorders are zymotic diseases. The word "*zymotic*" is derived from the Greek term "*zume*" (which signifies "*a ferment*").

873. The poison of infection is rendered harmless directly it is *diffused in pure fresh air*.

So soon as the infectious exhalation of this kind of zymotic disease is mingled with fresh pure air, it is, in the first place, immediately diluted, and then completely destroyed by the resolving power of the oxygen (865). It is probable that the poison of infection never produces any of its dire results in other living frames, when it has to pass through a distance of six feet, filled with fresh pure air, after leaving its source.

874. All that is necessary to prevent the spread of infectious diseases is *perfect ventilation*.

This is the one important practical fact to be thoroughly understood, and constantly borne in mind. Infection is produced by an unusually impure condition in the blood of living animals. Nature's plan of destroying infection is the mingling of it with pure air; therefore, let nature have fair play, do not cross its plans, and it will at once arrest the evil. All the fearful and deadly infectious diseases that visit the human race, sweep away their hosts of victims, because these live in close, badly ventilated, and too often crowded dwellings, and because they are of uncleanly habits. The first thing to do in any case of infectious disease is to open all the doors and windows, and to keep the sufferer very clean both in person and linen. When these measures can be perfectly carried out, the infectious character of the disorder very soon vanishes. Infectious disease results simply and entirely from the poisonous exhalations of the disordered body being received in a confined atmosphere, where they are concentrated by accumulation, and where there is not a corresponding sufficiency of oxygen for their quick destruction.

875. Fevers *generally tend to their own cure.*

The principal treatment that is necessary in disorders of this kind is to let nature have fair play. If the purest air is supplied, the most perfect cleanliness is scrupulously observed, and nothing is given or done that can continue the oppression of the system, nature soon rights itself. The excreting organs, step by step, expel the offending matter from the blood, and the balance of health is gradually restored. The low diet, the quiet, and the other valuable measures directed by medical science, all have the object in view, in the first place, of removing or withholding such influence as would oppose nature in its beneficent work; and, in the second place, of aiding its operations, so far as this can be done by artificial appliances. Every one ought to understand this, in order that he may be prepared to yield intelligent and efficient obedience to medical suggestions and directions, in case of being placed in positions in which such may be required.

876. Some zymotic diseases are produced by subtile aerial poisons, which, however, *do not seem to have issued from the bodies* of people affected by the same disease. Such diseases are communicated from some subtile poison that is widely diffused in the air. There can be no doubt of this, because many persons are seen to be attacked with the disorder at once, who have had nothing else in common but the breathing of the same atmosphere; but these persons do not give the disease to their neighbors. The disorder is literally caught of the air, and not of surrounding people. In these cases, the poison, when it is breathed, acts as a ferment in the blood, producing derangement and disorder amongst its constituents; but the formation of other subtile poison exactly like itself, out of some of the constituents, is not a part of the derangement. When yeast is added to a solution of pure sugar, it causes it to ferment, and changes it into alcohol, but it does not then lead to the formation of fresh yeast capable of producing further fermentation, as it does when it is placed in the glutinous solution of malt. The influence of these subtile ferments in producing disease without renewing themselves as poisons, is, when compared with the influence of true infection, exactly like the power of yeast over a solution of pure sugar, as compared with its power over a solution of glutinous malt (800, 871).

877. When zymotic diseases fall, as if from the air, *upon great numbers of people at once,* instead of being infectiously communicated from one person to another, they are termed *epidemics.* The word "*epidemic*" is derived from the Greek terms " *epi* " and " *demos* " (which signify " *upon* " and " *the people*"). Infectious disorders *creep along* slowly,

and step by step, because they are communicated from person to person, and because each body, after being affected, must brew its own poison before it can send this on to its neighbor; but epidemic diseases burst out in large districts, and affect great numbers of people simultaneously. *Typhus fever* is an *infectious* disease, and creeps slowly from house to house: but *influenza* is *epidemic;* it falls at once on thousands, as if out of the air. All London was affected with influenza at once on the two days of April the 3d and 4th, in the year 1833.

878. Epidemic poison is *even more subtile* than infectious poison.

Its presence is ascertained by its results, but both its source and its nature are alike unknown. Possibly it may be a half organic complex poison, like that of infection: but it *is not at once destroyed in the same way, by free exposure to the open atmosphere:* free exposure rather increases than diminishes the chance of the disorder being caught. But, on the other hand, it always runs its course, and brings itself to an end in about five or six weeks. Whereas infectious disease may pass on from house to house, year after year, if some means are not taken to arrest its progress. Infectious diseases are at the present time more under human control than epidemics. Many persons believe that the poison of epidemy is of a vegetable rather than of an animal character, and many that it even consists of living fungi of extremely minute dimensions floating in the air. If this be so, still there can be no doubt that the sudden appearance of those fungi is due to some unusually impure condition of the atmosphere, which is capable of serving them as food.

879. Some zymotic diseases are *at once epidemic and infectious.*

That is, they are produced by some subtile mischief floating in the air amongst a great number of people at once; but they are then capable of being communicated from person to person, under favoring circumstances,

just as typhus fever would : this is the case with cholera.
The poison of *epidemic cholera* travels from country to
country through the air, until it has completely encom-
passed the earth ; but it does not produce a real outbreak of the disease until, in its progress, it sweeps over
some place that is peculiarly fitted for the concentration
and reproduction of the poison. It is with cholera much
the same as it is with a spark, which leads to no explosion until it *falls on gunpowder.* The gunpowder contains the slumbering material of the explosion, and the
spark is the force which wakes that slumbering material
into activity. The gunpowder of the choleraic spark is
filth. So soon as the subtile epidemic effluvium of
cholera floats into a town, where it finds the air laden
with impure animal exhalations—that is, with complex
half organic poisons, such as have been already described,
it mingles with these, and, by a touch, converts them into
an agent of the most fearful energy. When in its
highest intensity, the poison of cholera extinguishes animal life almost as immediately as the vapor of prussic
acid (862). So soon as an explosion of cholera poison
has been produced in some fatal spot, it spreads all
round, as waves spread round the spot in which a stone
has been dropped into smooth water, growing, however,
constantly thinner and less active as it spreads. In this
way it at last gets so weak as to exhaust itself, unless it
finds by accident, during its progress, new food for its
virulence, when in that spot another explosion happens,
a fresh outbreak occurs, and the renewed poison again
spreads all round, weakening as it goes. This is how it
happens that cholera is at once epidemic and infectious.
The poison is epidemic so long as it floats through tolerably pure air, but it creates a fearful infection so soon
as it mingles with the vapors of putrefying matters. The
epidemic influence is like the spark, and the impure vapors are like latent gunpowder, possibly harmless in
itself, but dangerous in the extreme when sparks are
near. If the epidemic influence and the impure vapors
come together, an explosion of infection takes place, and

9*

the disease appears in great violence. When individuals who are not living in very confined or impure localities suffer from cholera, it is because the impurities are already in their own systems in great force. There is some peculiar condition present in them that predisposes them to suffer from the injurious influence, although it only comes to them as gently as epidemic. The extreme probability is that the exhalations of putrid animal matters are very near to the nature of cholera poison, but not quite it. In order to become so, they only require some slight internal change to be made amidst their atoms, and they then at once are converted into poisonous compounds of fearful energy, something of the nature of prussic acid vapor. The influence which pushes on the impure vapors into the state of active cholera poison seems to be the floating to them of a slight breath of the active poison as an epidemy from a distance through the air, in connection with the presence of moisture and warmth. This then acts upon them as a sort of ferment.

880. Mixed zymotic diseases that become infectious under peculiar circumstances, *may be checked by the same precautionary measures as ordinary* infectious diseases (874).

Hence the necessity, when any formidable epidemic of the nature of cholera, visits the land, of at once removing all decaying matters from the neighborhood of human dwellings, in order that when the epidemic breath floats over them, it may not find there the material it is capable of converting into infection. He, who at such times lives in the midst of filth, is, literally, like one standing over gunpowder, and waiting the arrival of the spark that is to blow him to destruction. Cholera only visits severely thickly-peopled places, in which decaying organic matters are so abundant that it is very difficult to get them removed as fast as they are produced. Towns can only be saved from the ravages of cholera, by perfecting arrangements for the cleansing and venti-

lating of their houses and streets. An abundant supply
of pure water and fresh air, and the immediate removal
of all decomposing products that are invariably formed
by the operations of animal life, are absolutely essential
to the preservation of health. Hence, when. men con-
gregate in social communities for the advantages of mutual co-operation and companionship, artificial means
must be devised to ensure these blessings to the community. Science has no more important or beneficent
duty to perform than that it fulfils when employed in
this service. Man cannot, at present, stop the blast of
the epidemic,—but he can render it faint, and nearly
powerless, by depriving it of its food. He has been
unable to fathom the mysteries of its origin, or to check
its career, but he nevertheless can prevent it from being
nursed into a pestilence. When every man understands
his own wondrous organization, and has learned how it
is affected by external influences, towns will cease to be
devastated at intervals by pestilential disease. Typhus
fever, cholera, and other similar visitations, now sweep
away their thousands of victims from the earth, because the great majority of mankind is ignorant in matters that it is of the utmost importance every rational
creature should know.

881. Some zymotic diseases are neither
epidemic nor infectious, but are *confined to
certain spots* on the earth.

In these spots such diseases are found constantly during particular seasons of the year. They are, in fact,
not caught either of the atmosphere (speaking in a wide
sense) nor of affected persons, but *of the spot* where
they occur. They do not fall suddenly upon numbers
and sweep on over the surface of the earth, neither do
they spread from person to person; but *they seize on all
who come to dwell in their haunts.*

882 Uninfectious zymotic disorders, that
prevail only in particular confined localities,
are termed *endemic diseases.*

That is, diseases which occur *in a people* of a particular district. The word endemic is derived from the Greek terms "*en*" and "*demos*" (which signify "*in the people*"). *Ague* is an endemic disease. It occurs to the inhabitants of certain marshy places, but it never communicated from one person to another like typhus fever; neither does it travel from one spot to another like influenza or cholera.

883. The poison that produces endemic disease is a subtile half-organic substance, produced *from the decay of vegetable matters*.

It is called by the very expressive name of *malaria*, which is merely the Italian term for "*bad air.*" Endemic diseases only occur in moist, marshy places, during warm seasons. Some physicians doubt whether the malarious poison is really a product of the decay of vegetable matter; but they are not able to say what else it is. At any rate, it is invariably connected with the presence of such decaying matter, with moisture and with warmth; for, whenever deep marshes are thoroughly drained, and whenever the temperature of the air falls below 60 degrees, all endemic disease disappears. Thus much at least is certainly known regarding the poison that produces intermitting fevers. It is an effluvium or vapor poured out from the surface of the earth of certain spots into the air, when that earth is saturated with stagnant water, and when the temperature is above 60 degrees. Almost always the moist earth contains, too, a great abundance of mouldering vegetable matter. In temperate climates, malarious poison does not commonly acquire any very great intensity; but, in the hotter regions of the earth, it sometimes becomes of equal virulence to any other organic poison known. The dreadful scourge of the coast of Africa, and of the West Indies, known as *the yellow fever*, is an uninfectious endemic disorder, produced by malarious poison. There are some interesting peculiarities in this subtile

agent that ought to be known to every one, who is likely ever to visit its haunts. It is more dangerous at
night than in the day time; it lurks principally near
the surface of the ground, especially if there be any mist
there; it drifts along in the direction of the wind; it is
absorbed and destroyed when it passes over the surface
of any reservoir of water; it clings to the foliage of
bushes and trees; it is most abundantly produced on
the instant when moist places are just dried up, and its
powers are invariably diminished by the operations of
cultivation. These facts point with sufficient clearness
to the precautions that chance visitors ought to take in
malarious places, such as the fens of Lincolnshire, the
Pontine marshes near Rome, and the banks of the Niger.
These precautions are—never sleep at night on the
ground; keep as much as possible on the opposite side
of pieces of water to that from which the wind blows;
avoid trees and shady spots, and also the borders of
swamps which are most affected by changes from wet to
dry, and the reverse; and keep as much as possible
within the boundaries of cultivated land.

884. *Really healthy and vigorous* frames
can be exposed to the influence of most of the
zymotic poisons with impunity.
Strong and healthy men can breathe even the poison
of typhus fever and cholera for a very long time, with
out taking the disease, provided that poison be not in its
highest degree of intensity. It is thrown out from the
system as rapidly as it is received, and without having
set up any destructive operations in the blood.

885. In a general way, there must be a
predisposition to suffer any particular disease, as well as exposure to the influence that
is capable of exciting it, before that disease
appears.
This predisposition is in itself really a disordered
state. The blood is already in some way impure and

deranged, although not in a sufficient degree to produce any very marked amount of discomfort, until a fresh cause of disturbance is added. This is seen in what has already been stated regarding the influence of cold (780). Cold does not take any particularly injurious effect upon the system, unless its powers of resistance and re-action have been already sapped by exhaustion and fatigue; then the exhaustion constitutes the predisposition, and the chill is the additional cause which completes the mischief. It is like the last hair which breaks the camel's back. The same thing is illustrated in the course of all epidemics. It is the weak and the already disordered alone that fall before their attack.

886. Predisposition to disease is almost always contracted *through perversity or ignorance.*

People who suffer from disease, nearly always do so, either because they have been in the habit of eating or drinking too much; because they have breathed impure air; because they have lived indolent lives; because they have not sufficiently attended to the laws of cleanliness; because mentally or bodily they have worked too hard; or, because they have indulged themselves in vicious sensual enjoyments.

887. Disease is hence a penalty *that may be avoided* through wisdom and the exercise of self-control.

If intemperance, want of cleanliness and free ventilation, laziness, immoderate and unregulated exertion, and vicious pursuits, all of which are circumstances within the compass of human control, were banished from the world, nine-tenths of the diseases that now afflict the human race, and abridge the lives of numbers, would be avoided. Many suffer disease because they have inherited predisposition from their parents; but this makes no real difference in the case, for then this predisposition is the result of some mistaken or wilful proceeding of the parent: it is indeed "the sin of the father visited upon the child;" and it has been beneficently ar-1anged that even such inherited predispositions ultimately disappear when better habits are observed. All this makes it apparent how essential it is that every one should be sent into the world with a clear understanding of what the general structure of his body is, and what the duties are that he owes to himself. A naturally weak, but wise and temperate man, has far greater chances of steering himself clear of the rocks and shoals of life, and of reaching to honorable length of days, than the strongest man who knows nothing about the structure of his wonderful frame, and allows every unregulated impulse and desire to sway him as it arises.

888. Sensual temptations were designed *to be struggled against,* and not to be yielded to (616–618).

And this for the most obvious reason, that in every case, strength can only be developed through exertion and conflict. The blacksmith's arm only grows brawny after he has labored long at the anvil. So, in the same way, man's moral nature only gets developed and strengthened through exerting itself. The skilful sea-man is not trained by sailing about in a pleasure-barge upon smooth water, but in struggling with the fury and the danger of the tempest upon the open sea: and so also the most earnest and admirable men are such as have struggled amidst the difficulties and dangers of life, and have come out from the struggle victorious over their own instincts a .d passions. There is this broad differ-ence between an existence of sensual indulgence, and one of refined intelligence: the man who leads the former, enjoys present gratification, at the cost of having to pay in the future a heavy fine of pains and penalties, which must alike detract from his own happiness and abridge his share of usefulness; whilst in the latter there is even higher enjoyment in the present, and this without the certainty of there being consequent punishment to come.

The pleasures of reason and intellect, which are, after all, the most unqualified delights in themselves that man can secure, entail no subsequent hours of sad and bitter repentance; no ultimate experience of conscious degradation; no consequent throbs of an exhausted and diseased body, that can never again be restored to the integrity from which wilful and perverse misconduct, or scarcely less lamentable ignorance, has cast it down.

889. Disease is an *uncomfortable or painful state*, resulting from the embarrassed and discordant action of the vital organs of the frame.

Health is the natural, harmonious, and either unconscious, or agreeable action of the various organs of the living body. It is characterized by the absence of all painful and uneasy sensations, and results from the presence of sound structure and constitution in every part of the frame. Disease is the opposite of this: it is characterized as the name implies (*dis-ease*), by the presence of uneasy or painful sensations somewhere, and results from the presence of unsound structure or constitution of some part. It rarely happens that disease is unconscious, and this is a very beneficent and fortunate arrangement, for the painful or uncomfortable feelings that are its attendants, draw the attention to the mischief that is in progress, and suggest the search for measures of relief (628).

890. Disease is sometimes *confined to one organ or to one part* of the living frame.

Inflammation produced by external violence is an illustration of this. If a severe blow is inflicted upon some part of a healthy body, that part soon after grows red, swells, and becomes painful and hotter than usual. It is then said to be inflamed (the word being derived from the Latin term "*inflammo*," "*to set on fire*"). These changes of its state are caused in the following manner: first, more blood flows to it than it ordinarily

receives; but the increased quantity does not leave the
vessels of the injured part as freely as it enters them;
it stagnates in them and dilates the channels. The
dilated vessels try to contract, but being unable to do
so, they relax after the useless effort and enlarge themselves still more. The corpuscles of the stagnant blood
then adhere together in these dilated vessels, and constitute a sort of plug; its serum and fibrin, stopped by the
plug, are forced through their walls, and so a tenacious
mass, quite unlike the ordinary structure, is formed.
This mass is red, because there is more red blood in the
capillary vessels that enter into its composition than
usual. It is hot, on account of the increased quantity of
blood that comes to it. It is affected with swelling,
because a great deal of blood flows to, whilst very little
leaves, it; and it is painful because the increased bulk
squeezes the extremities of the sentient nerves lying in
connection with it. When a limited portion of the body
is inflamed, its structure is rendered unsound, and consequently becomes the seat of uneasy sensations: in other
words, that portion of the body is, for the time being,
diseased.

891. Disease is nearly always *connected
with some disordered condition of the blood*.

An individual organ of the frame cannot be diseased
without in some way producing disorder in the blood.
Thus, if some part be injured by external violence, and
be caused to inflame, as described above, it does not take
from the blood what it ought, and, besides this, its
diseased condition otherwise directly affects the stream
of the circulation. Amongst other things, the colorless
corpuscles (346) and the fibrin (362) get to be proportionally much increased. As every organ is essential
to the perfection of the body, the structure of no one of
the group can be injured without the entire balance of
the frame being in consequence disturbed. It is in this
way that the bad management of some one organ (868)
leads to general disorder. Fever mostly commences

26

with some one particular structure ceasing to do its
usual work. Then the blood becomes affected in consequence of its fault, and this affected blood carries
derangement to all other parts.

892. Nature *always attempts to cure disease,*
and very commonly succeeds, if allowed fair
play.

The history of disease abounds with the most wonderful contrivances adopted by unaided nature for the
removal of disorder. The natural cure of inflammation
is one striking instance of these reparative contrivances.
When any part of an otherwise healthy body has been
inflamed by the application of external violence, as
described above (890), if the part is kept quiet, and the
person who has suffered the injury, refrains from the
use of stimulating food, the stagnant mass of adherent
corpuscles and effused fibrin soon gets hollowed out by
new channels, the blood begins again to move onwards
through these, and step by step the obstruction grows
less and less, until it entirely disappears. So, in the
same way, when the general system is affected with
fever, if the sufferer remains quietly in bed, and lives
very sparingly, the excreting organs that have been oppressed and inactive, in a few days resume their labors,
and then begin to make up for lost time, by expelling
the offending matters that are redundant in the blood;
and so, as comparatively little of any thing has been
added to it during the time, the blood soon gets into a
sufficiently pure state for the ordinary condition of health
to be restored.

893. Medicines *assist nature in her cura-
tive processes* when they are judiciously ad-
ministered.

The skilful physician, when any person suffering
disease is placed under his care, knows exactly what
nature is trying to do for the removal of the mischief,
as also that certain medicinal agents, introduced into

the frame, will greatly assist its proceedings, and sometimes even enable them to be carried through, where
they could not without such artificial assistance. Medicines operate in two different ways; sometimes they
merely increase or alter the action of some particular
organ or organs, and so lead to an influence being produced on the blood through their agency; at other times,
they get themselves taken at once into the blood, and
change the condition of its entire mass, and therefore
also of every structure into which it is carried. Rational
persons, who are suffering from disease, are far more
likely to benefit from skilful medical advice, if they
fully understand its purport, because then they take
care always to be in the strictest accord with it. The
more intelligent a patient is, the greater are the physician's chances of being enabled to guide him right in
his pursuit of health.

894. Natural decay results *from the exhaustion of cell vitality* (267).

In early life, successive generations of living vesicles
come into being more quickly than the old ones are destroyed; consequently, then, an accumulation of them
takes place, and the body increases in size. During
middle age, the production and the destruction of fresh
generations of vesicles keep even pace with each other,
and therefore the body maintains pretty much the same
bulk: in advanced life, vesicles are destroyed more
quickly than they are produced, and so the body wastes
year after year. This result arrives, because the power
of cell production that was handed down from the one
parent cell of the body (261) begins to fail. Languor
first falls upon it, but that languor then grows more and
more, until at length there comes a time when the reproductive force is altogether exhausted: no more new
cells are moulded as the old ones burst and dissolve
away. But when cell production ends, nourishment can
no longer be liquefied and absorbed into the blood for its
renovation (296); no more purification of its stream can
be effected (447); no more muscular contraction can be

produced (491); no more nerve influence or sensation can be originated (528). The consequence is, that all the ordinary operations of the wondrous economy cease, life is exchanged for death, and the organized structure is crumbled into dust.

895. Disease sometimes leads to *premature death*.

Man always dies prematurely when he does not attain to his appointed three-score years and ten. It sometimes happens that structure gets so much spoiled by disease, in organs whose integrity is essential to the continuance of life, that neither the reparative powers of nature, nor the most skilful appliances of art are able to restore them to their previous useful state; then, as a matter of course, life ends before its ordinary term.

896. When disease kills prematurely, it always does so by *arresting the movements of the circulating blood*.

The primary cause of the movement of the vital stream is the contractile action of the heart (384). This contractile action is kept up simply by the stimulant power of the red arterial blood upon the irritable muscular fibres of the left side of the organ (498). When, at any time, the arterial blood gets to be impure, the heart either fails or labors in its action, under the want of its ordinary stimulus. If the arterial blood gets very depraved indeed, the action of the heart stops altogether. and then, as no more blood is pumped onwards to the system, every organ within it is starved.

897. Death, induced prematurely by disease, *sometimes begins in the blood*.

This happens when from any cause a sufficient quantity of arterial blood is not formed to keep up the heart's play. *Death from starvation*, and from such disorders of the digestive apparatus as prevent the due supply of nutrition, occurs in this way.

898. Premature death *sometimes begins in the heart.*

That is, its own walls may become so faint and languid that they are not able to answer, even when properly stimulated by arterial blood. *Death from heart-faintness* occasionally happens when the organ itself is exhausted or injured from structural change.

899. Premature death sometimes *begins in the lungs.*

When the structure of the lungs gets so altered that the blood can no longer be arterialized as it passes through their capillaries, dull venous blood is sent on to the left side of the heart instead of bright arterial blood (424); but dull venous blood cannot stimulate the left side of the heart, so that the organ may keep up its play (896). *Death from suffocation* happens whenever the lungs are prevented from performing their proper office (709).

900. Premature death *sometimes begins in the brain and spinal cord.*

The mechanical movements which bring fresh air into the air cells of the lungs continually (703), and so affect the due arterialization of the blood, are kept up by the influence of the spinal cord (554); and many other actions, which are essential to the continuance of life, are ruled over and arranged by the sensorium (560). When the structure of the sensorium is so injured that these actions are not looked after, death soon ensues, or when that of the spinal cord is so spoiled that the movements of respiration cease. suffocation follows as certainly as if either the lungs were destroyed, or the vivifying air was shut out from their cavities. *Death from stupor* occurs whenever the great central masses of the nervous apparatus are paralyzed.

INDEX

THE END.